Lattice Methods for
Multiple Integration

Lattice Methods for Multiple Integration

I. H. SLOAN

Professor of Mathematics
The University of New South Wales

and

S. JOE

Lecturer in Mathematics
The University of Waikato

CLARENDON PRESS · OXFORD

1994

Oxford University Press, Walton Street, Oxford OX2 6DP

Oxford New York
Athens Auckland Bangkok Bombay
Calcutta Cape Town Dar es Salaam Delhi
Florence Hong Kong Istanbul Karachi
Kuala Lumpur Madras Madrid Melbourne
Mexico City Nairobi Paris Singapore
Taipei Tokyo Toronto
and associated companies in
Berlin Ibadan

Oxford is a trade mark of Oxford University Press

Published in the United States by
Oxford University Press Inc., New York

A catalogue record for this book is available from the British Library

Library of Congress Cataloging in Publication Data
(Data available)

ISBN 0 19 853472 8

Typeset by the author

Printed in Great Britain by
Bookcraft (Bath) Ltd
Midsomer Norton, Avon

Preface

Numerical integration in one dimension has been studied for centuries—
think of the rectangle, Simpson, and Gauss rules. However, numerical
integration for dimensions much greater than one had to wait for the advent
of the computer age to see significant progress. This is because of the 'curse
of dimensionality': the number of quadrature points rises alarmingly as the
number of dimensions increases.

The early history of numerical multiple integration, admirably summarized by Stroud (1971), saw most effort devoted to problems of small
or moderate dimension. Though low-dimensional problems remain important, there are areas of application (for instance atomic physics, quantum
chemistry, statistical mechanics, and Bayesian statistics) in which there are
many variables or degrees of freedom. Thus there is a demand for cost-
effective integration rules in six or ten or twenty (or even a few hundred!)
dimensions. Even with the powerful computers of today, high-dimensional
integrals can challenge available computer resources.

The high-dimensional problems arising in practice may have many fea-
tures that make life difficult. The integrand may contain singularities or
discontinuities; or, if smooth, may be concentrated in a small part of the
region, or may oscillate rapidly. The behaviour of the integrand may be
unknown: the integrand may be too complicated to understand mathematically, and the dimensionality too high to allow effective numerical
exploration. The geometry of the integration region may be difficult and
the region may be bounded or unbounded. And many methods, including favourites such as the product-Gauss rules, yield just one number at
the end of a calculation involving perhaps hundreds of thousands of function evaluations, so giving an approximate integral whose accuracy may be
impossible to determine.

For so rich a class of problems there can be no universal 'best' method.
Our interest in this book is in a very special class of rules known as 'lattice
rules'. Their virtue lies in their simplicity and in the fact that they can be
found in a practical way even when the dimensionality is high. Their theory
is elegant and attractive. But they have limitations: the integrand needs
to be reasonably smooth, the integration region just the s-dimensional unit
cube, and the integrand one-periodic in each of the s variables. (However,
we shall see that the class of problems is widened considerably if we are
prepared to apply an initial nonlinear transformation.)

Lattice rules are, in effect, generalizations of the humble rectangle rule
for an interval. Perhaps surprisingly, the higher-dimensional generaliza-

tions of the rectangle rule are very rich. (The reader may wish to sneak a glance at Figure 4.4 on page 79 to see the quadrature points for an interesting two-dimensional lattice rule.)

The rectangle rule is not generally considered one of the success stories of numerical analysis, but in fact if the integrand is smooth and one-periodic then the rectangle rule (which because of the periodicity is equivalent to the trapezoidal rule) is known to be highly effective (see Section 2.4). Even if the integrand is not naturally periodic (for example, through having different values at the two ends of the interval), it is possible to transform the problem into one with a periodic and somewhat smooth integrand by an appropriate nonlinear transformation, after which the rectangle rule should again be reasonably satisfactory. This philosophy also underlies the higher-dimensional generalization to lattice rules: we may be prepared to do some preliminary work to make the integrand periodic and somewhat smooth in return for being able to achieve reasonable accuracy with an extremely simple approximation rule.

Lattice rules began life in the late 1950s and the 1960s through the efforts of some distinguished number theorists, including E. Hlawka, N.M. Korobov, S.K. Zaremba, and L.K. Hua. The early history has been well summarized by Niederreiter (1978a). The subject has been rejuvenated over the past decade through the recognition of a much more general notion of lattice rules. Having been privileged to play some part in this rejuvenation, we shall be well pleased if we can impart to the reader not only the end results, but also some of the flavour and excitement of this elegant yet practical subject.

This book is aimed not only at graduate students of numerical analysis and applied mathematics, but also at practical scientists and others who have high-dimensional integrals which they need to approximate. Very little direct knowledge is assumed: although at times aspects of group theory, number theory, and Fourier analysis enter the discussion, we have tried wherever possible to make the treatment self-contained. We have also tried to provide a guide for those who may wish to skip the theory and move as quickly as possible to the applicable methods. But we would also hope the book will be attractive to mathematicians, because of the graceful and useful fusion of number theory, algebra, and analysis provided by the topic of lattice rules.

In writing this book we have received helpful comments and criticisms from many people, and especially from H. Niederreiter, R. Cools, L.N. Trefethen and J.N. Lyness. To the authors alone, however, belongs the responsibility for selection, accuracy, and interpretation of the material. In particular, we believe that the algorithms given in this book are correct, though we do not guarantee that they will meet your requirements for any particular application.

We thank the Association for Computing Machinery for the right to use a paper which appeared in their journal ACM Transactions on Mathematical Software. This paper (Joe and Sloan 1993) is the basis for some of the material in Chapter 10.

We would also wish to record our deep gratitude to the Australian Research Council for its sustained support, which has enabled much research on lattice methods to be carried out over the past decade. The fruits of this research form a major part of the contents of this book.

Finally, we acknowledge gratefully the patient support of our families, who have allowed us to spend so much of our lives bewitched by lattice rules.

Contents

1

Introduction

1.1 Numerical multiple integration

This book is concerned with a class of methods, called lattice methods, for the numerical evaluation of multiple integrals. Multiple integrals occur in a great variety of practical applications. Very often there are just two or three integration variables (in which case the integral is said to be two-dimensional or three-dimensional). For low-dimensional problems of this kind, the traditional methods summarized later in this chapter are generally satisfactory and lattice methods probably have little to offer. But increasingly there is a demand for cost-effective methods for high-dimensional integrals. Such problems can arise whenever there are many degrees of freedom—as, for example, in the quantum mechanics of atoms, molecules, or nuclei, or the theory of probability. The methods to be discussed in this book may provide a practical tool for integrals of dimension up to say 10 or 20, or perhaps even higher for suitable problems.

We shall be introduced to lattice rules (of which the prototype is the one-dimensional rectangle rule) in the next chapter. First, however, the problem needs to be specified, after which, in order to set the scene, we shall briefly review the more traditional approaches to numerical multiple integration. Shortcuts for busy readers are indicated in Section 2.1.

For simplicity, we shall mostly assume that the problem is formulated as an integral over the s-dimensional unit cube $C^s = [0,1]^s$, that is,

$$If = \int_{C^s} f(\mathbf{x})\, d\mathbf{x} = \int_0^1 \cdots \int_0^1 f(x_1, \ldots, x_s)\, dx_1 \cdots dx_s. \qquad (1.1)$$

(The exception is in Section 9.3, where we look briefly at problems over unbounded regions.) This is of course a limitation, but not as great a one as might appear at first sight. Any two-dimensional integral which can be written in the form

$$\int_0^1 \int_{u(t_2)}^{v(t_2)} g(t_1, t_2)\, dt_1\, dt_2$$

can be transformed into the desired form by the elementary change of variable

$$x_1 = \frac{t_1 - u(t_2)}{v(t_2) - u(t_2)}, \qquad x_2 = t_2.$$

The same idea is easily extended to repeated integrals in any number of dimensions.

In almost all problems met in practice, f is continuous and indeed even analytic, except for possible singularities (in the sense of points of non-analyticity) situated at isolated points or on lower-dimensional surfaces. Numerical integration rules that rely on point evaluations cannot be expected to handle infinite singularities, and their performance is usually significantly degraded by any form of singularity that lies in the interior. It is generally good advice to determine the position and nature of all singularities, and either to remove or weaken them by an appropriate nonlinear transformation, or to break the region into pieces so that all singularities lie on the boundary. Some rules (such as the product-Gauss rules to be discussed below) are less troubled by boundary singularities than interior singularities. In the case of lattice methods any singularities on the boundary will be further weakened by nonlinear transformations of the kind to be discussed in Section 2.12.

For the one-dimensional integral

$$If = \int_0^1 f(x)\,dx,$$

there are many quadrature rules of the form

$$Qf = \sum_{j=0}^{N-1} \omega_j f(x_j),$$

differing in the choice of sets of real numbers $\{x_0, \ldots, x_{N-1}\}$ (the 'quadrature points' or 'abscissae') and $\{\omega_0, \ldots, \omega_{N-1}\}$ (the 'weights' or 'coefficients'). Almost invariably the quadrature points are required to lie in $[0,1]$, and there is a strong preference for rules in which all the weights are positive. The N-point Gauss rule, a popular favourite, has all weights positive and integrates exactly all polynomials of degree $\leqslant 2N-1$. For this and other aspects of one-dimensional quadrature see, for example, Davis and Rabinowitz (1984) or Engels (1980).

Similarly, the s-dimensional integral (1.1) may be approximated by a variety of quadrature rules of the form

$$Qf = \sum_{j=0}^{N-1} \omega_j f(\mathbf{x}_j), \tag{1.2}$$

with $\mathbf{x}_0, \ldots, \mathbf{x}_{N-1}$ belonging to C^s. (In the literature, multidimensional integration rules are often called 'cubature' rules to distinguish them from the one-dimensional 'quadrature' rules. We shall not make that distinction in this book.) The number of distinct quadrature points is the 'order' of the quadrature rule.

1.2   Product rules

The most obvious approach to approximating the s-dimensional integral (1.1) is to view it as a repeated integral involving s separate one-dimensional integrals. It is then natural to apply a one-dimensional rule, say

$$\int_0^1 g(t)\,dt \approx \sum_{j=0}^{n-1} \mu_j g(t_j),$$

to each one-dimensional integral in the repeated integral (1.1), in this way obtaining the 'product rule'

$$Qf = \sum_{j_s=0}^{n-1} \cdots \sum_{j_1=0}^{n-1} \mu_{j_1} \cdots \mu_{j_s} f\left(t_{j_1},\ldots,t_{j_s}\right). \tag{1.3}$$

Product rules are easy to use, since the set of quadrature points is just the s-fold product of the set $\{t_0,\ldots,t_{n-1}\}$, and the weights are just the corresponding products of the weights. They are conceptually simple, with an error analysis that follows immediately from that for the underlying one-dimensional rule.

Product rules have much to recommend them if the dimensionality s is small. Accepted wisdom is that the product-Gauss rule is the preferred choice in all dimensions s up to 4 or 5.

For large values of s, however, product rules become unmanageable: the order of the rule (1.3) is

$$N = n^s,$$

which for fixed n grows alarmingly as s increases. Thus in five dimensions any product ten-point rule has $N = 10^5$; and for $s = 20$ even a product two-point rule gives $N = 2^{20} \approx 10^6$.

As a variant on product rules of the form (1.3), it is possible to have product rules in which not all the one-dimensional rules for each of the s variables are the same. For further details, see Stroud (1971).

1.3   Rules of prescribed polynomial degree

Most rules for approximating the integral (1.1) are designed to integrate a certain class of polynomials exactly. The rule (1.2) is said to be of polynomial degree d if it integrates exactly all polynomials of the form

$$\sum_{j_1+\cdots+j_s \leqslant d} a_{j_1\ldots j_s} x_1^{j_1} \cdots x_s^{j_s},$$

but is not exact for at least one monomial of degree $d + 1$.

The s-fold product of the one-dimensional n-point Gauss rule integrates exactly every monomial of the form

$$x_1^{j_1} \cdots x_s^{j_s} \quad \text{for } 0 \leqslant j_i \leqslant 2n - 1, \quad 1 \leqslant i \leqslant s,$$

but not the monomial x_1^{2n}, and so is of polynomial degree $2n - 1$.

By allowing rules that are not product rules, it is usually possible to find rules that are more efficient than the product-Gauss rules, in the sense of having polynomial degree $\geqslant 2n - 1$, yet requiring fewer quadrature points than the n^s points of the product-Gauss rule.

For a contemporary survey of methods for constructing rules of prescribed polynomial degree, see Cools (1992). Such rules exist in profusion, especially for small values of s. For a recent compilation, see Cools and Rabinowitz (1993). The topic lies outside the scope of this book, because lattice rules are constructed by a quite different principle. The next topic brings us closer to lattice rules.

1.4   Rules of prescribed trigonometric degree

As an alternative to integration rules of prescribed algebraic polynomial degree d, it is possible to obtain integration rules of prescribed trigonometric degree d. A trigonometric monomial in the variables $x_1, \ldots, x_s$ is a function of the form

$$e^{2\pi \imath h_1 x_1} e^{2\pi \imath h_2 x_2} \cdots e^{2\pi \imath h_s x_s}, \tag{1.4}$$

where $\imath^2 = -1$ and $h_1, \ldots, h_s \in \mathbb{Z}$. Here $\mathbb{Z}$ is the set of integers. The degree of the monomial (1.4) is $\sum_{j=1}^{s} |h_j|$. Now let $\mathbb{T}^s$ denote the space of trigonometric polynomials in s variables, that is, the set of all finite linear combinations of trigonometric monomials. The degree of a trigonometric polynomial is the maximum of the degrees of the monomials from which it is composed. The subspace of $\mathbb{T}^s$ that we are interested in is $\mathbb{T}_d^s$, the space of all trigonometric polynomials of degree at most d.

The rule (1.2) is said to be of trigonometric degree d if $Qf = If$ for all $f \in \mathbb{T}_d^s$ and if there exists a $g \in \mathbb{T}_{d+1}^s$ such that $Qg \neq Ig$. It may be shown (Mysovskikh 1987) that the dimension of $\mathbb{T}_d^s$ is given by

$$\dim(\mathbb{T}_d^s) = \sum_{j=0}^{n} \binom{n}{j} \binom{d}{j} 2^j.$$

Since both the points and weights in (1.2) may be chosen freely, it may be possible to obtain an integration rule of trigonometric degree d by taking

N to be less than $\dim(\mathbb{T}_d^s)$. However, it is known (see, for example, Cools and Sloan 1993) that a lower bound for N such that (1.2) has trigonometric degree d is $\dim(\mathbb{T}_{\lfloor \frac{d}{2} \rfloor}^s)$. A larger lower bound is known if d is odd.

These bounds lead to the notion of minimal quadrature rules of trigonometric degree, that is, for given d, quadrature rules for which the number of points N is as small as possible. Work in this area has been done by Mysovskikh (1987) and Cools and Sloan (1993). The latter paper displays minimal quadrature rules of trigonometric degree 1, 2, and 3 for arbitrary dimensions, as well as minimal quadrature rules of arbitrary trigonometric degree for $s = 1$ and $s = 2$. It turns out that some (but not all) of these integration rules are lattice rules. This relationship between such minimal quadrature rules and lattice rules has been studied in more detail by Beckers and Cools (1993).

1.5 Monte Carlo method

In the classical Monte Carlo method (Metropolis and Ulam 1949, Hammersley and Handscomb 1964) points $\mathbf{x}_0, \ldots, \mathbf{x}_{N-1}$ are chosen randomly in $[0, 1]^s$ and the integral (1.1) is then estimated by

$$Q_N f = \frac{1}{N} \sum_{j=0}^{N-1} f(\mathbf{x}_j). \tag{1.5}$$

Like the whole subject of gambling, the Monte Carlo method invites strong passions. Its virtues are considerable. Convergence is guaranteed (almost surely) by the central limit theorem, even if f is merely continuous. Moreover, the rate of convergence is independent of the dimensionality: the error $|Q_N f - If|$ is estimated by

$$\frac{\sigma(f)}{\sqrt{N}}, \tag{1.6}$$

where

$$\sigma^2(f) = \int_{C^s} f^2(\mathbf{x}) \, d\mathbf{x} - (If)^2$$

is the variance of f. Finally, and most importantly, a free estimate of the error is available: the variance may be estimated by

$$Q_N f^2 - (Q_N f)^2,$$

and its square root then used in (1.6) to estimate the error.

The negative aspects of the Monte Carlo method begin with the fact that truly random number sequences are not available: rather, one must make do with 'pseudo-random' sequences generated by a deterministic algorithm. Of course no deterministic sequence can be truly random, but it is possible to find 'random number generators' (of course misnamed) whose output appears random when subjected to simple statistical tests. For an entertaining discussion of pseudo-random sequences, see Knuth (1973). Recent developments in pseudo-random sequences are reviewed by Niederreiter (1992b). A typical random number generator is incorporated in the NAG routine G05CAF (Numerical Algorithms Group 1991)

Next, the rate of convergence reflected in (1.6), even if surprisingly good for high-dimensional problems, is still very poor. In practice, Monte Carlo methods rarely give better than 'experimental accuracy', say 1%. And to obtain another decimal digit of accuracy, the estimate (1.6) tells us that N must be increased a hundredfold.

But the aspect of the Monte Carlo method that causes most concern derives from the fact that in applications one almost invariably wants not just one integral, but a family of integrals with one or more continuously varying parameters. The statistical fluctuations associated with the classical Monte Carlo method (if it were feasible at all!) make it unsuitable for exploring integrals with continuously varying parameters.

There are many modifications and ramifications of the Monte Carlo method beyond those touched on here. For example, implementations of the Monte Carlo method often include some sort of 'variance reduction' technique. Such techniques include stratified sampling, importance sampling, and correlated sampling. Some brief details on stratified sampling are given in Section 1.7. For more detailed discussion on this and other variance reduction techniques, see Davis and Rabinowitz (1984), Hammersley and Handscomb (1964), or Kalos and Whitlock (1986).

1.6   Quasi-Monte Carlo methods

We have said that the random numbers theoretically required by the Monte Carlo methods are not in fact available, and that instead pseudo-random number sequences, generated by some deterministic algorithm, are used. In the quasi-Monte Carlo methods that apparent disability is turned on its head: one uses again the equal-weight formula (1.5) but chooses the sequence $\mathbf{x}_0, \ldots, \mathbf{x}_{N-1}$ to be 'better than random'.

The subject of quasi-Monte Carlo methods, thoroughly reviewed in the recent book by Niederreiter (1992b), can usefully be split into two branches.

In the first, just as in the classical Monte Carlo, the quadrature points $\mathbf{x}_0, \mathbf{x}_1, \ldots$ form an infinite sequence which is independent of N. Thus to increase N from one value to another, all that is needed is to calculate function values at the appropriate additional quadrature points. It seems reasonable to refer to methods of this type as 'open' quasi-Monte Carlo methods. It is a remarkable fact that specific infinite sequences $\mathbf{x}_0, \mathbf{x}_1, \ldots$ are known that are 'better than random', in the sense that the error in the rule (1.5) formed from the first N members of the sequence converges to zero more rapidly than the $O(N^{-1/2})$ rate of the classical Monte Carlo method, provided f is sufficiently well behaved. For example, the Halton sequence (Halton 1960) gives an error bound of the form

$$|Q_N f - If| \leqslant C \frac{(\log N)^s}{N} V(f), \tag{1.7}$$

where C is some constant independent of f and $V(f)$ is the variation of f in the sense of Hardy and Krause. In simple terms, $V(f)$ is a measure of the variability of f over the domain; a brief discussion may be found in Section 4.2, and more detailed information in Niederreiter (1978a).

An informal description of the Halton sequence is as follows: we write the counting numbers $0, 1, 2, \ldots$ in base 2,

$$0, 1_2, 10_2, 11_2, 100_2, 101_2, 110_2, \ldots,$$

then in base 3,

$$0, 1_3, 2_3, 10_3, 11_3, 12_3, 20_3, \ldots,$$

then in base 5, and so on through the first s primes. Then each of these expressions for the counting numbers is 'reflected in the origin' to yield a number in $[0, 1)$, that is, the digits in the original expression are reversed and written down after the decimal point. So, for example, the counting numbers in base 2 are reflected to yield

$$0, 0.1_2, 0.01_2, 0.11_2, 0.001_2, 0.101_2, 0.011_2, \ldots,$$

while those in base 3 become

$$0, 0.1_3, 0.2_3, 0.01_3, 0.11_3, 0.21_3, 0.02_3, \ldots.$$

This procedure yields, for each prime base, a sequence of numbers commonly referred to as a van der Corput sequence in that base. If $p(s)$ is the sth prime, then the Halton sequence is

$$
\begin{aligned}
\mathbf{x}_0 &= (0, \ldots, 0), \\
\mathbf{x}_1 &= (0.1_2, 0.1_3, \ldots, 0.1_{p(s)}), \\
\mathbf{x}_2 &= (0.01_2, 0.2_3, 0.2_5, \ldots, 0.2_{p(s)}), \\
\mathbf{x}_3 &= (0.11_2, 0.01_3, 0.3_5, \ldots, 0.3_{p(s)}), \\
&\ \vdots
\end{aligned}
$$

with the ith component of $\mathbf{x}_j$ being the number j in the base of the ith prime after 'reflection in the origin'. For a more formal description see Niederreiter (1978a, 1992b).

The estimate (1.7) is established by showing that the Halton sequence is well distributed over the unit cube, in the sense that the 'discrepancy' of its first N members is of order $O(N^{-1}(\log N)^s)$. (For a definition of discrepancy see the references cited above.) According to Niederreiter (1992b, p.33) it is widely believed that this order of convergence is the best possible with respect to order. On the other hand, the Halton sequence is certainly not the best possible with respect to the implied constant. For a discussion of sequences with better constants see the chapter on (t,s) sequences in Niederreiter (1992b).

In the second type of quasi-Monte Carlo method (which we might refer to as 'closed' methods) the quadrature points depend on N, and so would be better written as $\mathbf{x}_0^{(N)}, \ldots, \mathbf{x}_{N-1}^{(N)}$. Lattice rules are notable examples of closed quasi-Monte Carlo methods. We shall see in subsequent chapters that the freedom to choose all quadrature points anew for each value of N allows lattice rules to yield a better rate of convergence than can be achieved by any open quasi-Monte Carlo method, if f is suitably well behaved.

But even if f is not well behaved it is possible to improve the rate of convergence marginally over the order shown in (1.7) if we are prepared to use a closed quasi-Monte Carlo method. Hammersley (1960) proposed a point set with N points in which the first component is taken evenly spaced and the remaining components contain the components of the leading members of the Halton sequence moved one place to the right. Thus the Hammersley point set is

$$
\begin{aligned}
\mathbf{x}_0^{(N)} &= (0, \ldots, 0), \\
\mathbf{x}_1^{(N)} &= \left(\frac{1}{N}, 0.1_2, 0.1_3, \ldots, 0.1_{p(s-1)} \right), \\
\mathbf{x}_2^{(N)} &= \left(\frac{2}{N}, 0.01_2, 0.2_3, \ldots, 0.2_{p(s-1)} \right), \\
&\quad \vdots \\
\mathbf{x}_{N-1}^{(N)} &= \left(1 - \frac{1}{N}, \ldots \right).
\end{aligned}
$$

With the Hammersley sequence the discrepancy is improved (Niederreiter 1978a, 1992b), and the estimate (1.7) is replaced by

$$
|Q_N f - If| \leqslant C \frac{(\log N)^{s-1}}{N} V(f).
$$

That is, there is now one less power of $\log N$ than before. (The constant C in these two bounds, and in other bounds throughout this book, may take different values.)

1.7 Adaptive methods

This is a large topic, to which we cannot do proper justice in this short space. All we can do is to introduce the essential ideas.

Adaptive methods are usually built upon four things: a quadrature rule for estimating the integral over subregions of specified geometry; a method of estimating the error in that estimate of the integral; an error acceptance criterion (local or global); and a subdivision strategy.

If the error criterion is local, then an estimate of the integral over a subregion is accepted whenever the error estimate for that subregion (or the error per unit volume) reaches a predetermined level. In contrast, with the generally preferred global acceptance criterion no subregion is accepted until a satisfactory value is reached for the aggregate of the error estimates over all subregions.

For a multidimensional adaptive package the most interesting questions, perhaps, are the choice of the geometry for the subregion (for example, simplices or rectangular solids) and the subdivision strategy. Simplices (that is, triangles or tetrahedra) can work well in two or three dimensions, as in the two-dimensional packages TRIADA of Haegemans (1977), CUB-TRI of Laurie (1982), TRIEX of de Doncker and Robinson (1984a,b), and DCUTRI of Berntsen and Espelid (1992), and the three-dimensional package DCUTET of Berntsen *et al.* (1993). Rectangular solids are another popular choice for the subregions and we shall look at this choice in more detail.

How should a rectangular solid be divided into subregions if the error for that subregion (whichever error criterion is used) is unacceptable? The obvious idea, no doubt, is to divide the region into 2^s smaller regions in which each solid is congruent to the original, but with all lengths scaled by a factor of $\frac{1}{2}$. The trouble with this strategy is that it leads to an explosion in the size of the problem when s is large. To overcome this difficulty, Van Dooren and De Ridder (1976) proposed a more modest subdivision, with the rectangular solid simply cut into two pieces by slicing along one of the mid-planes, with the decision as to which of the s possible axes to favour hinging on which of the axes has the largest fourth divided difference. This subdivision strategy is a key part of the software package ADAPT of Genz and Malik (1980). In that package two quadrature rules are used on each subregion, one of polynomial degree 5 and the other of polynomial degree 7. The absolute difference between the two estimates on each subregion is used as an error estimate, and at each stage the subregion with the largest error estimate is bisected. As before, this bisection is perpendicular to the direction in which the local fourth divided difference of f is largest. A more complicated error estimate is used in the package DCUHRE, a successor to ADAPT. Further details on DCUHRE may be found in Berntsen *et al.* (1991a,b).

In the package ADAPT described in the preceding paragraph the underlying quadrature rules are rules of certain polynomial degree. A different idea lies behind the adaptive Monte Carlo routine D01GBF in the NAG software library. This routine is based on the work of Lautrup (1971), and uses the Monte Carlo method with stratified sampling. In stratified sampling the original cube C^s is subdivided into subcubes and a basic Monte Carlo integration is applied to each subcube. At the start of the process, these subcubes have equal volume. During each iteration an estimate of the integral and variance on each subcube is obtained. Summing these quantities then gives an estimate of the overall integral and the variance. It is then possible to see the contribution to the overall variance from each of the subcubes. This information is used to perform a new subdivision along each axis for the next iteration so that there is a reduction in the variance and an approximate doubling of the number of subcubes.

Now we turn to the real subject of this book: lattice rules.

2
Lattice rules

2.1 Shortcuts for busy readers

A reader whose principal interest lies in applications can turn at any time to
the algorithmic development in Chapter 10. Important practical material
on the nonlinear transformations usually needed to push typical integrands
into a suitable form for the application of lattice rules is given in Section
2.12. Key theoretical definitions are in Sections 2.6–2.8 and 3.2. And even
the busiest reader is advised to undertake the following brief tour of lattice
rules. Section 2.3 sets out the plan of this book.

2.2 A brief tour of lattice rules

Lattice rules are quadrature rules designed to approximate the integral
(1.1) of f over the unit s-dimensional cube, in situations in which f is
reasonably smooth and also one-periodic with respect to each component
of $\mathbf{x}$; that is,

$$f(\mathbf{x}) = f(\mathbf{x} + \mathbf{z}) \quad \text{for all } \mathbf{z} \in \mathbb{Z}^s \text{ and all } \mathbf{x} \in \mathbb{R}^s. \tag{2.1}$$

(As we shall explore more fully in Section 2.12, preliminary transformations
of the given integral may be needed to bring the problem into this form.)

Lattice rules, as we have said, are generalizations of the one-dimensional
rectangle rule

$$R_n f = \frac{1}{n} \sum_{j=0}^{n-1} f\left(\frac{j}{n}\right). \tag{2.2}$$

Note that, because of the periodicity assumption, we might equally have
said generalizations of the one-dimensional trapezoidal rule

$$T_n f = \frac{1}{2n} f(0) + \frac{1}{n} \sum_{j=1}^{n-1} f\left(\frac{j}{n}\right) + \frac{1}{2n} f(1); \tag{2.3}$$

but for simplicity we shall continue to refer to the rectangle rule.

What is the generalization of the rectangle rule to higher dimensions? It turns out that there are many answers. Perhaps the most obvious is the product-rectangle rule given by

$$R_{n^s} f = \frac{1}{n^s} \sum_{j_s=0}^{n-1} \cdots \sum_{j_1=0}^{n-1} f\left(\frac{j_1}{n}, \ldots, \frac{j_s}{n}\right). \tag{2.4}$$

However, as this rule has $N = n^s$ quadrature points, it is not cost-effective if the dimensionality is high. The problem is that even if the rate of convergence seems satisfactory when expressed in terms of n, it can be very slow indeed when expressed in terms of N. For example, if $f(\mathbf{x}) = x_1^2(1 - x_1)^2$, a function of just one variable, then the rule (2.4) gives exactly the same result as the one-dimensional, n-point rectangle rule applied to $x^2(1 - x)^2$, for which the error is of order n^{-4}. Since the order of the rule (2.4) is $N = n^s$, this translates to a convergence rate of only $N^{-4/s}$ for that rule, which is poor indeed when s is large.

Our first interesting generalization of the rectangle rule is the 'method of good lattice points', originated by the number theorists Korobov (1959) and Hlawka (1962) and rediscovered by Conroy (1967), a physical chemist. In this method the approximation takes the form

$$Qf = \frac{1}{N} \sum_{j=0}^{N-1} f\left(\frac{j}{N}\mathbf{z}\right), \tag{2.5}$$

where N is the chosen number of quadrature points and $\mathbf{z}$ is a carefully selected integer vector. (Mathematically it does not matter that some of the points $j\mathbf{z}/N$ lie outside the cube C^s, since the periodicity property (2.1) means that an integer vector can be subtracted from the argument of f to bring the point back within C^s.) The method of good lattice points has an extensive literature. There are detailed reviews by Niederreiter (1978a, 1988, 1992b) and Haber (1970, 1983), with the former concentrating on the mathematical properties and the latter emphasizing numerical analysis aspects. A version of this method is implemented in the NAG mathematical software library routines D01GCF and D01GDF (Numerical Algorithms Group 1991).

We shall have much to say later about the order of convergence of the method of good lattice points, but for the moment let us make just one simple observation: if N is prime and z_1 is not a multiple of N, then the rule (2.5) applied to $f(\mathbf{x}) = x_1^2(1 - x_1)^2$ is equivalent to an N-point rectangle rule applied to the function $x^2(1 - x)^2$, because the first components of the argument of f (reduced modulo 1) are just the points of the N-point rectangle rule in a different order. Thus the error is of order N^{-4}, compared to $N^{-4/s}$ for the product-rectangle rule (2.4). When s is large the difference is dramatic!

More generally, whenever f is a function of just one of its variables the method of good lattice points with N prime and a sensible choice of $\mathbf{z}$ makes full use of N independent function evaluations, whereas the product-rectangle rule uses just $N^{1/s}$ of them.

The definition of a more general 'lattice rule' is prompted by the observation that the product-rectangle rule and the method of good lattice points are in some ways quite similar: both are equal-weight rules in which the quadrature points form a regular array, and both reduce in the one-dimensional case to the rectangle rule. In the modern sense a lattice rule, of which these two rules are examples, is any rule of the form

$$Qf = \frac{1}{N} \sum_{j=0}^{N-1} f(\mathbf{x}_j),$$

in which the points $\mathbf{x}_0, \ldots, \mathbf{x}_{N-1}$ are all the points of an 'integration lattice' L that lie in the half-open unit cube $U^s = [0, 1)^s$. An integration lattice is a discrete subset of $\mathbb{R}^s$ which is closed under addition and subtraction, and which contains all the integer vectors. We shall learn later how to classify, count, and construct lattice rules.

It appears that the first mention of the general lattice rule was made by Frolov (1977). Later this idea was rediscovered and developed systematically by Sloan and Kachoyan (1984, 1987) and Sloan (1985).

The basic properties of lattice rules are developed later in this chapter. We shall see simple examples of integration lattices and lattice rules, discover a quadrature error expression involving the Fourier coefficients of f, and meet the so-called 'dual lattice' of an integration lattice.

An important fact now known about lattice rules is that every such rule can be written as a multiple sum of the form

$$Qf = \frac{1}{n_1 n_2 \cdots n_t} \sum_{j_t=0}^{n_t-1} \cdots \sum_{j_1=0}^{n_1-1} f\left(\frac{j_1}{n_1}\mathbf{z}_1 + \cdots + \frac{j_t}{n_t}\mathbf{z}_t\right), \qquad (2.6)$$

where $\mathbf{z}_1, \ldots, \mathbf{z}_t$ are integer vectors. Conversely, every expression of this form is a lattice rule. Later we shall meet a classification scheme for lattice rules, due to Sloan and Lyness (1989), which is based on this representation, and which leads to the important concept of 'rank'. Put simply, the rank of a lattice rule is the minimum value of t required to write the lattice rule as an expression of the form (2.6).

We shall see that the rank of a lattice rule can take any value between 1 and s inclusive. The product-rectangle rule given by (2.4) is an example of a rule of the maximal rank s, whereas the rule (2.5) used in the method of good lattice points is obviously a rule of rank 1. Rules with every rank between 1 and s also exist (in fact there are very many such rules), but have been recognized and studied only in very recent times.

The practical implementation of lattice rules which we describe in Chapter 10 makes use of a finite sequence of lattice rules $Q_0, Q_1,\ldots,Q_s$, given by

$$Q_r f = \frac{1}{2^r m} \sum_{k_r=0}^{1} \cdots \sum_{k_1=0}^{1} \sum_{j=0}^{m-1} f\left(\frac{j}{m}\mathbf{z} + \frac{(k_1,\ldots,k_r,0,\ldots,0)}{2}\right), \qquad (2.7)$$

$0 \leqslant r \leqslant s$, with m odd and $\mathbf{z}$ a well-chosen integer vector. The first member of this sequence,

$$Q_0 f = \frac{1}{m} \sum_{j=0}^{m-1} f\left(\frac{j}{m}\mathbf{z}\right),$$

is a rule of rank 1, as in the method of good lattice points. The subsequent members of the sequence $Q_1, \ldots, Q_s$ have the important property that the quadrature points for each member include all of the points of the preceding member and an equal number of new points—the rules are 'embedded'. The final member of the sequence,

$$Q_s f = \frac{1}{2^s m} \sum_{k_s=0}^{1} \cdots \sum_{k_1=0}^{1} \sum_{j=0}^{m-1} f\left(\frac{j}{m}\mathbf{z} + \frac{(k_1,\ldots,k_s)}{2}\right), \qquad (2.8)$$

turns out to be a rule of the maximal rank s, and is the one recommended for estimating the integral.

There are some significant attractions to the rule Q_s given by (2.8). First and foremost, because $Q_s f$ is a member of an embedded sequence (and indeed of many different such sequences), it is possible to generate, without cost, an estimate of the error. Second, it turns out to be relatively cheap to find a good choice of $\mathbf{z}$ in (2.8) for a given number of quadrature points: the cost is only a fraction of that of finding a good choice of $\mathbf{z}$ in the formula for the method of good lattice points given by (2.5). Third, it can be shown theoretically (in a lengthy exercise culminating in Chapter 6) that relatively 'good' rules of the form (2.8) do, in fact, exist. The first step in this theoretical exercise is to show that $Q_s f$ can be rewritten as

$$Q_s f = \frac{1}{2^s m} \sum_{k_s=0}^{1} \cdots \sum_{k_1=0}^{1} \sum_{j=0}^{m-1} f\left(\frac{j}{2m}\mathbf{z} + \frac{(k_1,\ldots,k_s)}{2}\right).$$

In other words, $Q_s f$ is equivalent to a '2^s copy' of the rule Q_0, that is, to the rule obtained by dividing the unit cube into 2^s subcubes of edge length $\frac{1}{2}$, and then applying a scaled version of the rule Q_0 to each of these smaller cubes. The particular virtue of 2^s copies of rules of rank 1 will be a major theme later in this book.

2.3 The plan of this book

In the remainder of this chapter we shall develop the basic properties of lattice rules, starting from the example of the rectangle rule in one dimension. Chapter 3 is devoted to the classification of lattice rules, and, in particular, defines properly the important notion of the rank of a lattice rule. Chapter 4 is devoted to the rank-1 rules which underlie the method of good lattice points. It is here that we shall meet many ideas and techniques which can also be used for more general lattice rules. In particular, we meet an averaging technique from the number theorist's armoury: the idea is that in order to prove that there exists a rule with some desired property, it suffices to show that the average over some convenient family of rules has the required property, since it then follows that there must be at least one rule in this family whose performance is as good as average.

Chapter 5 begins the quantitative study of lattice rules of rank greater than 1. We establish some criteria for measuring the 'goodness' of general lattice rules, these being extensions of the criteria already introduced in Chapter 4 for rank-1 rules. A simple argument in Chapter 5 shows that higher-rank rules can perform as well as rank-1 rules from the point of view of order of convergence, but reveals nothing about the magnitude of the errors. As a first step toward a quantitative comparison, we describe the results of some numerical experiments on rank-2 lattice rules.

In Chapter 6 we study lattice rules of the maximal rank s more seriously. It will be shown that rules of this class have a very simple characterization: they are 'copies', in the same sense as in the example at the end of the preceding section, of rules of lower rank. It turns out that rules that are copies of rank-1 rules are almost as simple to study as rank-1 rules themselves. Importantly, we shall see theoretical and numerical evidence that the best of the rules of this class are competitive in accuracy with the best of the rank-1 rules.

Certain lattice rules of intermediate rank, generalizing (2.7), are considered in Chapter 7.

Lattice rules for nonperiodic integrands are covered in Chapter 8. The work described here is still in its infancy.

Some other topics in the theory of lattice rules are covered in Chapter 9. This chapter includes material on lattice rules for integrals where the integration region is $\mathbb{R}^s$.

In Chapter 10 we turn to practical aspects of lattice rule computation. The computational scheme presented there, introduced already at the end of the preceding section, rests on the fact that the last and best lattice rule in the finite sequence is a well-chosen rule which contains many embedded lattice rules of lower order, that is, rules which can be obtained by retaining just a subset of the quadrature points. This property is exploited so as to obtain, with negligible computational effort, an estimate of the error.

In more detail, the algorithm presented in Chapter 10 yields an estimate of the integral, an estimate of the error, and also a sequence of estimates of the integral from an embedded sequence of lattice rules which culminates with the chosen rule.

Chapter 11 concludes the book with a number of experimental comparisons with other methods, including Monte Carlo and adaptive methods.

2.4   The rectangle rule

The one-dimensional rectangle rule (2.2) is the prototype of all lattice rules, so we begin the study of lattice rules with a brief exploration of its properties.

The rectangle rule (2.2) converges to the exact integral If as $n \to \infty$ for all Riemann integrable functions, because $R_n f$ is a Riemann sum. On the other hand we know that, when there is no requirement for f to be one-periodic, $R_n f$ can be a relatively poor approximation to If. For example, for the integrand

$$f(x) = x, \quad 0 \leqslant x \leqslant 1,$$

it is easily seen, either graphically or by a direct calculation, that

$$R_n f = \frac{1}{2} - \frac{1}{2n},$$

so that

$$|R_n f - If| = \frac{1}{2n}.$$

Yet it is known that $R_n f$ (or, equivalently, the trapezoidal approximation $T_n f$ defined by (2.3)) can be a remarkably good approximation to If if f has a smooth one-periodic extension to the whole real line. Consider, for example, the choice

$$f(x) = e^{\sin 2\pi x}.$$

For this example we find the results shown in Table 2.1, given to the 16 decimal digit accuracy of the machine used to do the calculations.

Why does the rectangle rule work so well when the integrand is smooth and one-periodic? The usual explanation is by way of the Euler–Maclaurin expansion (Davis and Rabinowitz 1984, p.136). (In brief, the Euler–Maclaurin expansion is an asymptotic series in which the kth term is a known multiple of the difference between the $(2k-1)$th derivatives of f at the two ends of the interval. If f is smooth and one-periodic then the derivatives at the two ends of the interval are equal, and so all terms of the asymptotic expansion vanish.)

For our present purposes, however, we find it more useful to give an explanation in terms of Fourier series, because the same arguments will be useful later for much more general lattice rules. Let us assume that the

Table 2.1 Errors for the rectangle rule when applied to $f(x) = e^{\sin 2\pi x}$

n	$R_n f$	$R_n f - I f$
2	1.000000000000000	-0.266065877752008
4	1.271540317407622	0.005474439655614
8	1.266066076964489	0.000000199212481
16	1.266065877752009	0.000000000000001
∞	1.266065877752008	

integrand f of (1.1), when extended periodically to the whole line, has an absolutely convergent Fourier series. That is,

$$f(x) = \sum_{h=-\infty}^{\infty} \hat{f}(h)e^{2\pi i h x}, \quad -\infty < x < \infty, \tag{2.9}$$

where

$$\hat{f}(h) = \int_0^1 e^{-2\pi i h x} f(x)\,dx, \quad h \in \mathbb{Z},$$

and where the right-hand side of (2.9) is absolutely convergent, that is,

$$\sum_{h=-\infty}^{\infty} \left|\hat{f}(h)\right| < \infty. \tag{2.10}$$

We may observe that the Fourier series is also uniformly convergent, from which it follows that f is continuous (from the theorem that the uniform limit of a sequence of continuous functions is continuous). Because f is also one-periodic, it is a continuous function satisfying $f(0) = f(1)$. Thus the present argument can apply only to integrands f that have a natural continuous one-periodic extension.

To work out the effect of the rectangle rule on f, we can now apply it term by term to the Fourier series (2.9). (Since $R_n f$ is only a finite sum, there is no problem in changing the order of summation.) Thus we obtain

$$R_n f = \sum_{h=-\infty}^{\infty} \hat{f}(h) R_n\left(e^{2\pi i h x}\right).$$

Now the rectangle rule applied to $e^{2\pi i h x}$ yields just

$$R_n\left(e^{2\pi i h x}\right) = \frac{1}{n}\sum_{j=0}^{n-1}\left(e^{2\pi i h/n}\right)^j$$

$$= \begin{cases} 1, & \text{if } h \text{ is a multiple of } n, \\ 0, & \text{if } h \text{ is an integer not a multiple of } n, \end{cases} \tag{2.11}$$

where we have used nothing more than the high-school formula for the sum of a geometric series. Thus

$$R_n f = \sum_{k=-\infty}^{\infty} \hat{f}(kn).$$

Finally, since $If = \hat{f}(0)$, the error in applying the rectangle rule to f is

$$R_n f - If = \sum_{k=-\infty}^{\infty}{}' \hat{f}(kn), \tag{2.12}$$

where the prime on the sum, here and elsewhere, indicates that the zero term is to be omitted from the sum.

This error formula shows clearly why the rectangle rule works well for a smooth periodic function f: if the Fourier coefficients of f approach zero rapidly then the leading terms of the error, namely

$$R_n f - If = \hat{f}(n) + \hat{f}(-n) + \hat{f}(2n) + \hat{f}(-2n) + \cdots,$$

will become extremely small when n is large.

2.5   Some two-dimensional lattice rules

Lattice rules, as we have said already, are higher-dimensional generalizations of the rectangle rule. Before we come to the formal definitions in the next section, we explore here what we might reasonably mean by a two-dimensional generalization of the rectangle rule.

One obvious answer is the product-rectangle rule given by (2.4) with $s = 2$. But there are more interesting rules, which can lay equal claim to be generalizations of the rectangle rule. Consider this example of order 5:

$$Qf = \frac{1}{5}\left[f(0,0) + f\left(\frac{1}{5},\frac{2}{5}\right) + f\left(\frac{2}{5},\frac{4}{5}\right)\right.$$
$$\left. + f\left(\frac{3}{5},\frac{1}{5}\right) + f\left(\frac{4}{5},\frac{3}{5}\right)\right]. \tag{2.13}$$

Like the one-dimensional rectangle rule, it is an equal-weight rule, with one point at the origin. The five quadrature points of this rule, shown in Figure 2.1, are distributed in a way which seems to be very regular. The

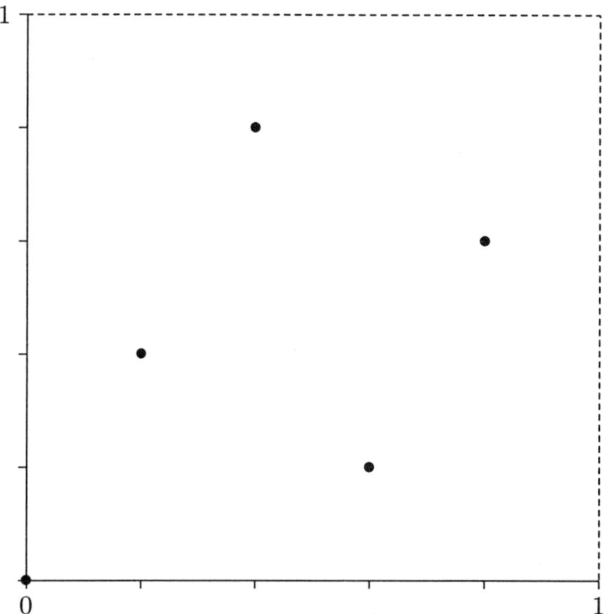

Fig. 2.1 The quadrature points for a two-dimensional lattice rule.

projections of the five points on to the y-axis, namely $\{0, \frac{2}{5}, \frac{4}{5}, \frac{1}{5}, \frac{3}{5}\}$, are just the five points of the one-dimensional rectangle rule. So, too, are the five projections on to the x-axis.

The rule (2.13) also shares one other important property with the rectangle rule, namely, that it admits a simple error analysis by way of Fourier series. In fact the rule (2.13) is our first significant example of a lattice rule, and it is a key property of all lattice rules that they permit a simple Fourier analysis of the error (see Section 2.9). We shall defer until later in this chapter a precise statement of the error in the rule (2.13), when we have had time to develop some machinery.

It is time now for us to turn to general lattice rules. But first, because a lattice rule is defined in terms of a lattice, we have to consider the question: what is a lattice?

2.6 Lattices

In mathematics the word 'lattice' has more than one meaning. For us in this book a lattice is defined this way.

Definition 2.1. *A lattice in* $\mathbb{R}^s$ *is a discrete subset of* $\mathbb{R}^s$ *which is closed under addition and subtraction.*

Notice that the point $\mathbf{0} = (0, 0, \ldots, 0)$ is in every lattice, since if $\mathbf{x}$ belongs to the lattice, so does $\mathbf{x} - \mathbf{x}$.

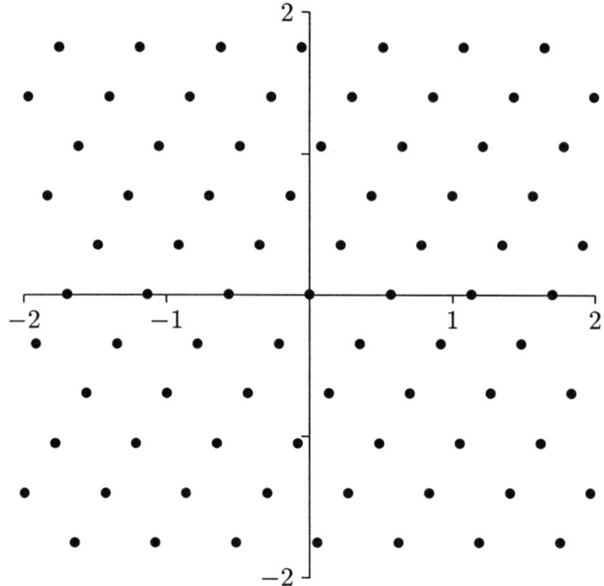

Fig. 2.2 A typical lattice in two dimensions.

An imaginary traveller in s-dimensional space would find a lattice very dull, since the view from every lattice point is the same. This is because a translation of the lattice by any point of the lattice leaves the lattice unchanged. It follows from this that every lattice, except for the lattice consisting only of **0**, is unbounded. A portion of a typical lattice in two dimensions is shown in Figure 2.2.

A lattice, though infinite, can be simply described in terms of a finite subset. We shall say that a linearly independent subset $\{\mathbf{g}_1, \ldots, \mathbf{g}_t\}$ is a 'basis' for the lattice if each point of the lattice is an integer linear combination of $\mathbf{g}_1, \ldots, \mathbf{g}_t$. The vectors $\mathbf{g}_1, \ldots, \mathbf{g}_t$ are 'generators' of the lattice. Every lattice has a basis, with $t \leqslant s$ (Cassels 1971, Section 3.4, Theorem VI). Of course the basis is not unique: for example, if $s \geqslant 2$ it is easily seen that the set of integer linear combinations is unchanged if we replace $\mathbf{g}_1$ by $\mathbf{g}_1' = \mathbf{g}_1 + \mathbf{g}_2$. Only if $t = s$ is the lattice truly s-dimensional. This is the only case of interest to us in this book.

Note that in the one-dimensional case there is just one generator to choose. It follows that in this case the only possible lattice is the set of all integer multiples of g, for g an arbitrarily chosen real number.

In a most natural way, a generator set $\{\mathbf{g}_1, \ldots, \mathbf{g}_s\}$ is associated with a 'fundamental parallelepiped', or 'unit cell',

$$\{\lambda_1\mathbf{g}_1 + \lambda_2\mathbf{g}_2 + \cdots + \lambda_s\mathbf{g}_s : 0 \leqslant \lambda_i \leqslant 1, \ 1 \leqslant i \leqslant s\}.$$

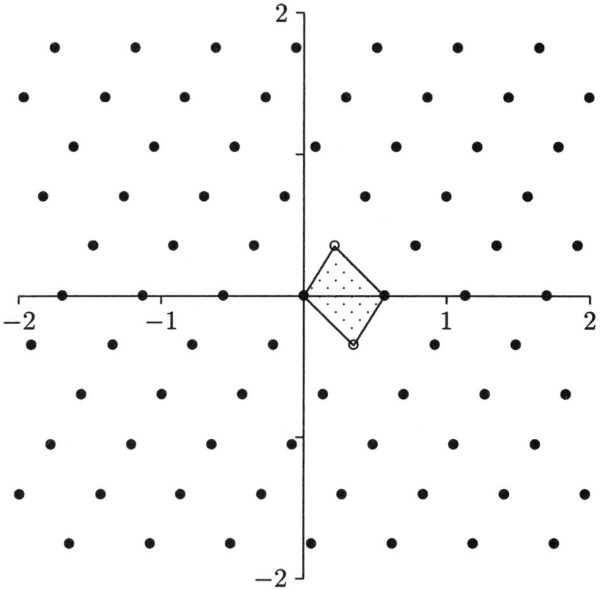

Fig. 2.3 A unit cell for the lattice shown in Figure 2.2.

In Figure 2.3 we show, for the lattice in Figure 2.2, one way of choosing the unit cell. The corresponding generators $\mathbf{g}_1$ and $\mathbf{g}_2$ are represented by the open circles.

Associated with the generator set $\{\mathbf{g}_1, \ldots, \mathbf{g}_s\}$ is a 'generator matrix' A, a square matrix whose rows are $\mathbf{g}_1^T, \ldots, \mathbf{g}_s^T$.

How can we measure the sparsity of a lattice? An obvious way is through the 'determinant' of the lattice, which is simply the volume of the unit cell. For a lattice L with generators $\mathbf{g}_1, \ldots, \mathbf{g}_s$, the determinant of L is

$$\det L = |\det A|.$$

The determinant is independent of the choice of generators (Cassels 1971, p.2). The average number of lattice points per unit volume is, of course, $1/\det L$.

Our intention, as indicated in Chapter 1, is to use the points of a suitable lattice as quadrature points for approximating the s-dimensional integral (1.1). We could conceivably use for this purpose any lattice which has a useful number of points in the integration region. Lattice points that lie outside the region would simply be ignored, and an appropriate equal weight (summing to 1) would be attached to those that remain. However, such an approximation would be rather crude. Note, for example, that if the scale of the lattice is steadily decreased, this quadrature approximation will generally change discontinuously whenever new quadrature points enter across the boundary. We shall do better than this by choosing the lattice more carefully. One price is that from now on the choice of integration region really matters.

2.7 Integration lattices

Throughout most of this book our integration region will be the s-dimensional unit cube,

$$C^s = [0,1]^s.$$

(The exception comes in Section 9.3, where we consider integration over unbounded regions.) Moreover, we will usually deal with integrands which (possibly after appropriate manipulation!) have the periodicity property (2.1). In that situation it turns out to be profitable to insist that the lattice has the same periodicity property. This leads us to the following definition.

Definition 2.2. *An integration lattice in $\mathbb{R}^s$ is a discrete subset of $\mathbb{R}^s$ which is closed under addition and subtraction, and which contains $\mathbb{Z}^s$ as a subset.*

The simplest integration lattice is, of course, just $\mathbb{Z}^s$ itself, the set of all s-vectors of integers.

Given any integration lattice L, another integration lattice can be obtained by scaling the lattice by a factor of $1/n$, that is, by dividing each point of the lattice by n, where n is any positive integer. In Figure 2.4 we show the integration lattice obtained in this way by scaling by $1/3$ the integer lattice $\mathbb{Z}^2$. The resulting lattice has generators $\mathbf{g}_1 = \left(\frac{1}{3},0\right)$, $\mathbf{g}_2 = \left(0,\frac{1}{3}\right)$, and determinant $1/9$.

Not all integration lattices are so uninteresting. Another integration lattice is that shown in Figure 2.5. One possible choice of generators for this lattice is

$$\mathbf{g}_1 = \left(\frac{1}{5},\frac{2}{5}\right), \qquad \mathbf{g}_2 = \left(\frac{2}{5},-\frac{1}{5}\right).$$

Another is

$$\mathbf{g}_1' = \left(\frac{1}{5},\frac{2}{5}\right), \qquad \mathbf{g}_2' = (0,1).$$

Both choices show the determinant to be $1/5$.

In the context of integration lattices, it is useful to define the 'half-open unit cube'

$$U^s := [0,1)^s.$$

(We use a half-open region to avoid double counting of lattice points lying on the boundary.) The number of points of an integration lattice L that

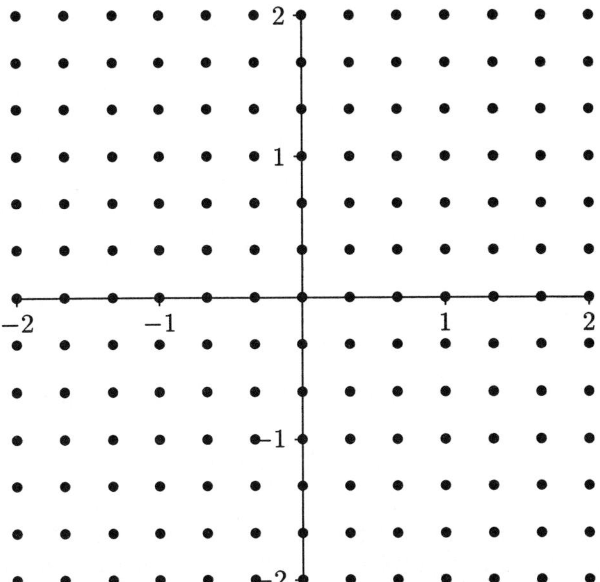

Fig. 2.4 The integration lattice obtained by scaling $\mathbb{Z}^2$ by $1/3$.

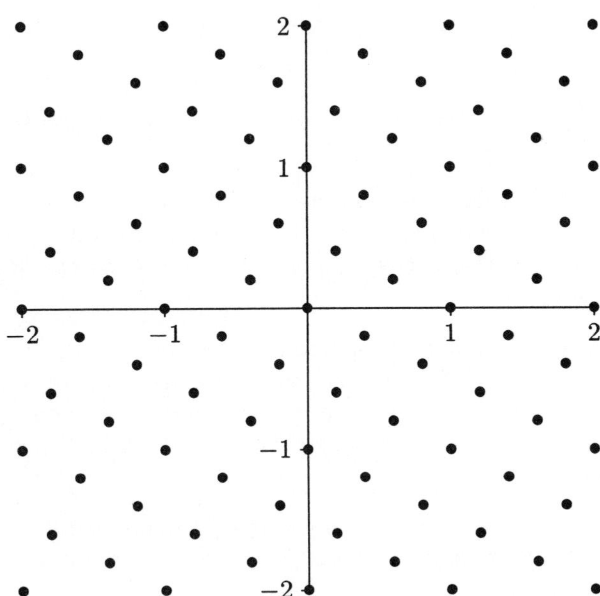

Fig. 2.5 Another example of an integration lattice.

lie in U^s, say N, is the average number of lattice points per unit volume, so that, from the definition of the determinant,

$$N = 1/\det L. \qquad (2.14)$$

For an integration lattice N is an integer; its values are 9 and 5 for the integration lattices in Figures 2.4 and 2.5 respectively. The lattice rule corresponding to the integration lattice L uses the points of the lattice lying in U^s as quadrature points. Thus the order of the corresponding lattice rule is N.

2.8 Lattice rules defined

To recapitulate, the problem we are considering is the approximate evaluation of

$$If = \int_{C^s} f(\mathbf{x}) \, d\mathbf{x}, \qquad (2.15)$$

where f is continuous (and preferably smooth) on the unit cube C^s, and f is also one-periodic with respect to each component of $\mathbf{x}$, so that (2.1) is satisfied.

Definition 2.3. *A lattice rule for* (2.15) *is a rule of the form*

$$Qf = \frac{1}{N} \sum_{j=0}^{N-1} f(\mathbf{x}_j), \qquad (2.16)$$

where $\{\mathbf{x}_0, \ldots, \mathbf{x}_{N-1}\}$ are all the points of an integration lattice $L \subset \mathbb{R}^s$ that lie in the half-open cube U^s.

In words, a lattice rule is an equal-weight rule whose quadrature points are all the points of an integration lattice that lie in the half-open unit cube.

The order N of the lattice rule (2.16) is related to the determinant of the lattice by (2.14).

Example 2.4. *For $s = 2$, the quadrature formula* (2.13) *given earlier is a lattice rule of order* 5. *The corresponding integration lattice is that in Figure 2.5, whose determinant is 1/5. The quadrature points, shown previously in Figure 2.1, are just the five points of the infinite lattice that lie in the half-open unit square.*

A lattice rule is entirely determined by specifying its integration lattice. The converse is also true: the integration lattice corresponding to the lattice rule (2.16) is given by

$$L = \{\mathbf{x}_j + \mathbf{z} : 0 \leqslant j \leqslant N - 1, \ \mathbf{z} \in \mathbb{Z}^s\}.$$

Example 2.5. *For $s = 2$,*

$$Qf = \frac{1}{9} \sum_{j=0}^{2} \sum_{k=0}^{2} f\left(\frac{j}{3}, \frac{k}{3}\right)$$

is a lattice rule of order 9 (the 'product-rectangle' rule). The corresponding integration lattice is shown in Figure 2.4.

Perhaps at this point it will become clear to the reader why we have already insisted on the periodic property (2.1) of f, even though the periodicity property is not used in the definition of the lattice rule. If f were not periodic then one might prefer, just as in one-dimensional quadrature, to use a 'trapezoidal-rule' version of the rule in Example 2.5, rather than the 'rectangle-rule' version that actually appears there. As long as we insist on the periodic property there is no distinction between the two, and no need to worry about the proper treatment of quadrature points on the boundary. In Chapter 8 we shall look at the application of the theory to the nonperiodic case. However, we want to emphasize that *lattice rules are really designed to take advantage of the periodicity and smoothness properties of f, and that their practical use for integrands which are not both smooth and periodic is generally not recommended.* The justification for this remark will become clear when we turn to the error analysis in the next section.

2.9 The lattice rule error

What lattice shall we use for our lattice rule? To answer that we need to understand the error in the lattice rule (2.16), for only then can we say whether one choice of integration lattice is better than another.

It turns out that the error in a lattice rule, just as for the rectangle rule, is easily expressed in terms of the Fourier components of f, provided f has a well-behaved periodic extension. This will pave the way for later consideration of the relative merit of different lattices.

We shall assume, in place of (2.9), that f has the absolutely convergent Fourier series representation

$$f(\mathbf{x}) = \sum_{\mathbf{h} \in \mathbb{Z}^s} \hat{f}(\mathbf{h}) e^{2\pi i \mathbf{h} \cdot \mathbf{x}}, \tag{2.17}$$

where $\mathbf{h} \cdot \mathbf{x} = h_1 x_1 + \cdots + h_s x_s$ is the ordinary inner product in $\mathbb{R}^s$, and

$$\hat{f}(\mathbf{h}) = \int_{C^s} e^{-2\pi i \mathbf{h} \cdot \mathbf{x}} f(\mathbf{x}) \, d\mathbf{x}, \quad \mathbf{h} \in \mathbb{Z}^s.$$

Now apply the lattice rule Q, given by (2.16), term by term to the series in (2.17), to obtain

$$Qf = \sum_{\mathbf{h} \in \mathbb{Z}^s} \hat{f}(\mathbf{h}) Q e^{2\pi i \mathbf{h} \cdot \mathbf{x}}. \tag{2.18}$$

The last step is to work out the effect of Q on $e^{2\pi i \mathbf{h} \cdot \mathbf{x}}$ for each $\mathbf{h} \in \mathbb{Z}^s$. It turns out that this has a very simple answer, in terms of the 'dual lattice' or 'reciprocal lattice' of the lattice associated with Q.

Definition 2.6. *Given a lattice L, the* dual lattice $L^{\perp}$ *is the set*

$$\{\mathbf{h} \in \mathbb{R}^s : \mathbf{h} \cdot \mathbf{x} \in \mathbb{Z} \text{ for all } \mathbf{x} \in L\}. \tag{2.19}$$

We shall have more to say in the next section about dual lattices. For the moment it suffices to observe that the dual of an integration lattice L is a subset of $\mathbb{Z}^s$: for the integration lattice by definition contains each unit vector $\mathbf{e}_i$, $1 \leqslant i \leqslant s$ (where $\mathbf{e}_i$ has a 1 in the ith component and zeros in all others), so that Definition 2.6 immediately tells us that if $\mathbf{h} \in L^{\perp}$, then $\mathbf{h} \cdot \mathbf{e}_i = h_i \in \mathbb{Z}$ for $1 \leqslant i \leqslant s$.

The following lemma is a natural generalization of the rectangle-rule result (2.11). It is taken from Sloan and Kachoyan (1987).

Lemma 2.7. *Let Q be the s-dimensional lattice rule (2.16), and let L be the associated integration lattice. Then for $\mathbf{h} \in \mathbb{Z}^s$,*

$$Qe^{2\pi i \mathbf{h} \cdot \mathbf{x}} = \begin{cases} 1, & \text{if } \mathbf{h} \in L^{\perp}, \\ 0, & \text{otherwise.} \end{cases}$$

Proof The first part is just as easy as in the one-dimensional case: if $\mathbf{h} \in L^{\perp}$ then $\mathbf{h} \cdot \mathbf{x}_j$ is an integer for each quadrature point $\mathbf{x}_j$, from which it follows that

$$Qe^{2\pi i \mathbf{h} \cdot \mathbf{x}} = \frac{1}{N} \sum_{j=0}^{N-1} e^{2\pi i \mathbf{h} \cdot \mathbf{x}_j} = \frac{1}{N} \sum_{j=0}^{N-1} 1 = 1.$$

The second part is rather deeper, but can be proved by the following argument, which has its genesis in an argument of Sobolev (1962) for numerical integration over the sphere. For each point $\mathbf{x}_j$, $0 \leqslant j \leqslant N-1$, we define a translation operator T_j, which operates only on functions with the periodicity property (2.1), by

$$T_j f(\mathbf{x}) = f(\mathbf{x} + \mathbf{x}_j).$$

It follows that

$$T_k T_j f(\mathbf{x}) = f(\mathbf{x} + \mathbf{x}_\ell) = T_\ell f(\mathbf{x}),$$

where $\mathbf{x}_\ell$ is the unique member of $\{\mathbf{x}_0, \ldots, \mathbf{x}_{N-1}\}$ that differs from $\mathbf{x}_j + \mathbf{x}_k$ by an integer vector.

Given a function f with the periodicity property (2.1), we define $\bar{f}$, the average of the translations $T_j f$, by

$$\bar{f} = \frac{1}{N} \sum_{j=0}^{N-1} T_j f.$$

Then $\bar{f}$ is invariant under translation, since

$$T_k \bar{f} = \frac{1}{N} \sum_{j=0}^{N-1} T_k T_j f = \frac{1}{N} \sum_{\ell=0}^{N-1} T_\ell f = \bar{f},$$

where we have used the fact that the translations $\{T_k T_0, \ldots, T_k T_{N-1}\}$ are just the translations $\{T_0, T_1, \ldots, T_{N-1}\}$ taken in a different order.

Now, if we apply the translation T_k to the function

$$\psi_{\mathbf{h}}(\mathbf{x}) = e^{2\pi \imath \mathbf{h} \cdot \mathbf{x}},$$

where $\mathbf{h} \in \mathbb{Z}^s$ and $\mathbf{x} \in \mathbb{R}^s$, then we find

$$T_k \psi_{\mathbf{h}} = e^{2\pi \imath \mathbf{h} \cdot \mathbf{x}_k} \psi_{\mathbf{h}}.$$

From this it is clear that if $\mathbf{h} \notin L^\perp$, then $\psi_{\mathbf{h}}$ is different from $T_k \psi_{\mathbf{h}}$ for some k in $0 \leqslant k \leqslant N - 1$. Also, we have

$$\bar{\psi}_{\mathbf{h}} = \frac{1}{N} \sum_{j=0}^{N-1} T_j \psi_{\mathbf{h}} = \left(\frac{1}{N} \sum_{j=0}^{N-1} e^{2\pi \imath \mathbf{h} \cdot \mathbf{x}_j} \right) \psi_{\mathbf{h}}.$$

Now, by the argument above, $\bar{\psi}_{\mathbf{h}}$ is invariant under the translation T_k, whereas for some k in $0 \leqslant k \leqslant N - 1$ $\psi_{\mathbf{h}}$ is not. The only way out is for both sides of the equation to vanish. Thus the expression in parentheses vanishes, completing the proof. ∎

Now we are in a position to state the s-dimensional generalization of (2.12). This result, first given by Sloan and Kachoyan (1987), is the main result of this section, and the foundation of all our subsequent error analysis.

Theorem 2.8. *Let Q be the s-dimensional lattice rule* (2.16), *and let L be the associated integration lattice. Moreover, assume that f has the absolutely convergent Fourier series* (2.17). *Then*

$$Qf - If = {\sum_{\mathbf{h} \in L^\perp}}' \hat{f}(\mathbf{h}). \tag{2.20}$$

Proof This follows immediately from (2.18) and Lemma 2.7, together with $If = \hat{f}(\mathbf{0})$. ∎

2.10   Shifted lattice rules

Sometimes it is useful to 'shift' the quadrature points of a lattice rule. One reason might be to ensure that no quadrature points lie on the boundary of the integration region. Another motivation for shifting the lattice will become more apparent when we consider, in Sections 4.6 and 10.3, an error estimation procedure based on a randomization technique.

For each $\mathbf{c} \in \mathbb{R}^s$, a 'shifted lattice rule' corresponding to the lattice rule (2.16) is

$$Q_\mathbf{c} f = \frac{1}{N} \sum_{j=0}^{N-1} f(\{\mathbf{x}_j + \mathbf{c}\}). \qquad (2.21)$$

Here, we have made use of the 'fractional' part of a vector:

Definition 2.9. *For any vector* $\mathbf{x}$*, let* $\{\mathbf{x}\}$ *be the vector in* U^s *obtained by taking the fractional part of each component of* $\mathbf{x}$*.*

It follows that $\{\mathbf{x}\} = \mathbf{x} - \mathbf{z}$, where $\mathbf{z}$ is an appropriate integer vector. Because of the periodicity assumption (2.1) we have $f(\{\mathbf{x}_j+\mathbf{c}\}) = f(\mathbf{x}_j+\mathbf{c})$, thus the $\{\}$ can always be omitted. We now give an expression for the error in the shifted lattice rule.

Theorem 2.10. *Let* $Q_\mathbf{c}$ *be the* s*-dimensional shifted lattice rule* (2.21), *and let* L *be the integration lattice associated with the (unshifted) lattice rule* (2.16). *Moreover, assume that* f *has the absolutely convergent Fourier series* (2.17). *Then*

$$Q_\mathbf{c} f - If = \sideset{}{'}\sum_{\mathbf{h} \in L^\perp} e^{2\pi\imath \mathbf{h} \cdot \mathbf{c}} \hat{f}(\mathbf{h}). \qquad (2.22)$$

Proof   From (2.17) and (2.21) we find

$$Q_\mathbf{c} f = \sum_{\mathbf{h} \in \mathbb{Z}^s} \hat{f}(\mathbf{h}) e^{2\pi\imath \mathbf{h} \cdot \mathbf{c}} Q e^{2\pi\imath \mathbf{h} \cdot \mathbf{x}}.$$

The result now follows immediately from Lemma 2.7.   ∎

2.11   The dual lattice

We have seen in Theorem 2.8 that the error for a lattice rule whose integration lattice is L is just the sum of the Fourier coefficients $\hat{f}(\mathbf{h})$ over all nonzero integer vectors $\mathbf{h}$ which belong to $L^\perp$, the dual of L. Thus the dual lattice is the key to an understanding of the error. In this section we give some examples of dual lattices, find a geometrical interpretation, and learn how to construct a generator matrix for the dual lattice.

In Figure 2.6 we show the dual of the lattice in Figure 2.4, the latter being the lattice obtained by scaling $\mathbb{Z}^2$ by $\frac{1}{3}$. The dual is of course just

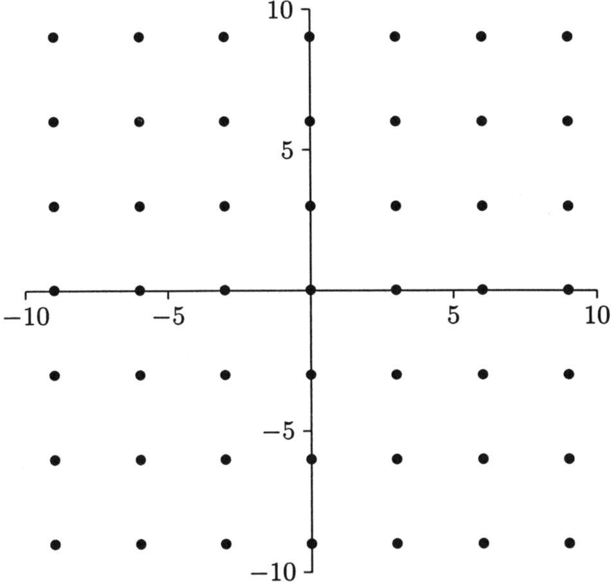

Fig. 2.6 Dual lattice for the lattice shown in Figure 2.4.

the integer lattice $\mathbb{Z}^2$ scaled by the reciprocal amount, namely 3. Figure 2.7 shows the dual of the lattice

$$L = \left\{ \left\{ j \left(\frac{1}{5}, \frac{2}{5} \right) \right\} + \mathbf{z} : 0 \leqslant j \leqslant 4, \ \mathbf{z} \in \mathbb{Z}^2 \right\}, \tag{2.23}$$

the lattice shown in Figure 2.5. For example, we easily verify that $(-1,3)\cdot\mathbf{x}$ is an integer for each point $\mathbf{x} \in L$, so that $(-1,3) \in L^\perp$. So, too, is $(-2,1)$ and hence so is any integer linear combination of $(-1,3)$ and $(-2,1)$.

Geometrically, the dual of the lattice L gives us information about the planes of dimension $s-1$ associated with L. For a given lattice, there are many different ways to draw families of equally spaced parallel planes in which each point of the lattice is on one of the planes, and each plane contains at least one point of the lattice. Each such system of planes corresponds to a point $\mathbf{h} \in L^\perp$, in that the equations of the planes can be written as

$$\mathbf{h} \cdot \mathbf{x} = \delta, \text{ for } \delta = 0, \pm 1, \pm 2, \ldots . \tag{2.24}$$

From elementary vector geometry, $\mathbf{h}$ is a vector normal to each of these planes, and the distance between each plane is $1/\|\mathbf{h}\|$, where

$$\|\mathbf{h}\| = (h_1^2 + \ldots + h_s^2)^{1/2}$$

is the Euclidean norm of $\mathbf{h}$.

As an example, in Figures 2.8 and 2.9 we show two such families of planes associated with the two-dimensional lattice (2.23). A dual lattice

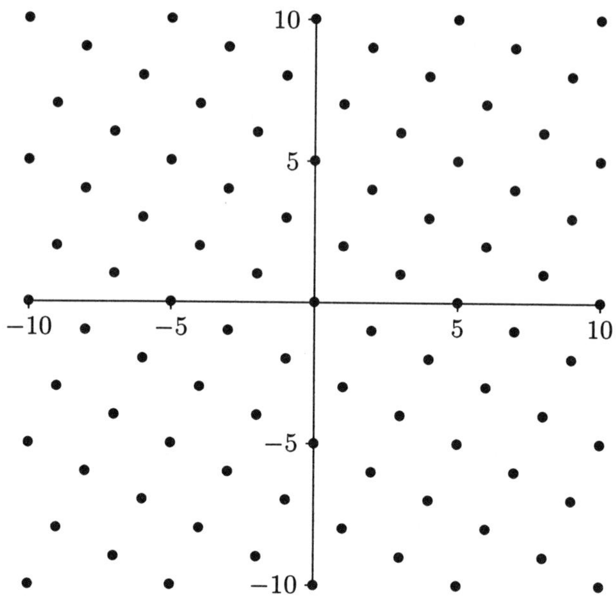

Fig. 2.7 Dual lattice for the lattice shown in Figure 2.5.

vector associated with the first is $\mathbf{h} = (-2, 1)$. The corresponding vector is
normal to the plane, and the distance between the planes is $\|(-2, 1)\|^{-1} = 1/\sqrt{5}$. Similarly, the second set of planes is associated with $\mathbf{h} = (-1, 3)$,
and the distance between the planes is $\|(-1, 3)\|^{-1} = 1/\sqrt{10}$.

There are of course infinitely many other such choices of $\mathbf{h}$, and, corre-
spondingly, infinitely many other families of planes. However, planes that
are close together are hard to see (look again at Figure 2.5!), because the
points in closely spaced planes must be sparsely distributed.

One may remark that this kind of thinking, in terms of the planes of
the lattice, will be very familiar to those who have studied the diffraction
of X-rays from crystals. In simple terms, one may think of the X-rays as
being reflected by families of planes of atoms. Thus it is natural to consider
the diffraction in terms of the dual lattice. The most important planes are
those for which the distance between them is large, because these are the
planes with the highest density of atoms.

If A is a generator matrix for L (see Section 2.6), we shall see that
$(A^{-1})^T$ is a generator matrix for $L^{\perp}$. For if we denote the rows of A by
$\mathbf{g}_1^T, \ldots, \mathbf{g}_s^T$ (so that $\{\mathbf{g}_1, \ldots, \mathbf{g}_s\}$ is a generator set for L), and denote the
columns of A^{-1} by $\mathbf{c}_1, \ldots, \mathbf{c}_s$, an arbitrary element $\mathbf{h}$ of $L^{\perp}$ may be written
as

$$
\begin{aligned}
\mathbf{h} &= A^{-1} A \mathbf{h} \\
&= \mathbf{c}_1 (\mathbf{g}_1 \cdot \mathbf{h}) + \cdots + \mathbf{c}_s (\mathbf{g}_s \cdot \mathbf{h}),
\end{aligned}
$$

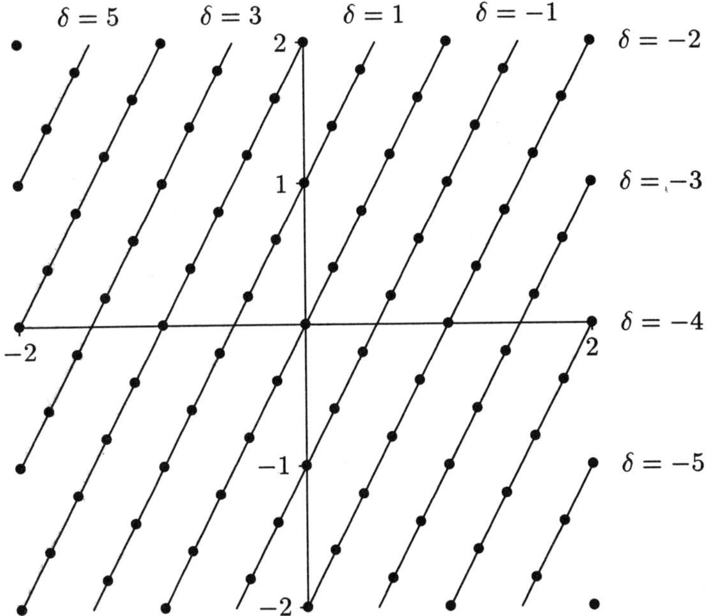

Fig. 2.8 The family of planes $(-2, 1) \cdot \mathbf{x} = \delta$.

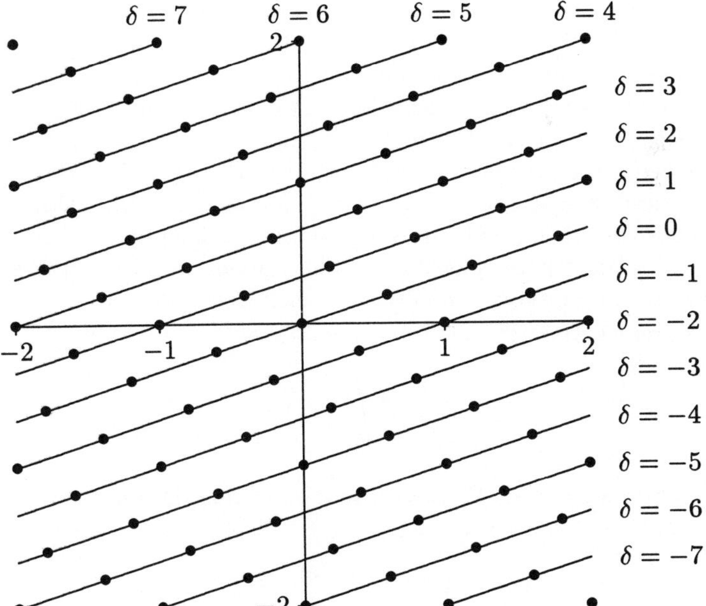

Fig. 2.9 The family of planes $(-1, 3) \cdot \mathbf{x} = \delta$.

which by Definition 2.6 is an integer linear combination of $\mathbf{c}_1, \ldots, \mathbf{c}_s$. Thus the columns of A^{-1} form a generator set for $L^{\perp}$, so $(A^{-1})^T$ is a generator matrix for $L^{\perp}$.

Before leaving the dual lattice, it may be useful to remind ourselves again of the significance of the dual lattice in the study of lattice rules. Lemma 2.7 tells us a remarkable fact: that every lattice rule integrates exactly the great majority of all the Fourier terms $e^{2\pi i \mathbf{h} \cdot \mathbf{x}}$. A given lattice rule fails only for those values of $\mathbf{h}$ which belong to the dual lattice. Disregarding the point $\mathbf{h} = \mathbf{0}$, because the constant function is integrated exactly, the dual lattice represents a graphic picture of failure. Looking again at, for example, the dual lattice in Figure 2.7, we see that the corresponding rule fails for (on average) a fraction of $1/N$ of the Fourier terms $e^{2\pi i \mathbf{h} \cdot \mathbf{x}}$. The most important failures are presumably for those values of $\mathbf{h}$ that are in some sense close to the origin, such as $\mathbf{h} = (-2, 1)$ and $\mathbf{h} = (-1, 3)$.

2.12   Periodizing the integrand

In the preceding sections we have emphasized that the integrand f in the integral (2.15) should be both continuous (and preferably smooth) and one-periodic with respect to each component of $\mathbf{x}$. Most integrands met in real life do not have all these properties: they may be continuous, and perhaps may even be naturally one-periodic with respect to some components, but rarely are they one-periodic with respect to every component of $\mathbf{x}$. Thus almost inevitably there is a need for some kind of preliminary transformation to force the integrand to be periodic.

Several different methods have been proposed for this purpose, but only one, which we might call the method of nonlinear transformations, seems to have been used for serious calculations. The problem of periodizing the integrand is discussed in detail by Zaremba (1972). More recently, numerical experiments with different strategies have been carried out by Beckers and Haegemans (1992a). In the present work we shall give most attention to the method of nonlinear transformations.

We consider first the one-dimensional case

$$If = \int_0^1 f(x)\, dx, \tag{2.25}$$

with f continuous on $[0, 1]$. In the method of nonlinear transformations, a transformation of the form

$$x = \phi(t), \tag{2.26}$$

with ϕ a smooth increasing function which maps $[0, 1]$ onto $[0, 1]$, changes the problem to one of the same form as (2.25) but with an integrand modified through the usual rules of calculus: the transformed integral is

$$If = \int_0^1 g(t)\,dt,$$

where

$$g(t) = f(\phi(t))\phi'(t), \quad 0 \leqslant t \leqslant 1. \tag{2.27}$$

In the s-dimensional case we just apply the same transformation to each component separately, to yield

$$If = Ig, \tag{2.28}$$

where

$$g(t_1,\ldots,t_s) = f\left(\phi(t_1),\ldots,\phi(t_s)\right)\phi'(t_1)\phi'(t_2)\cdots\phi'(t_s). \tag{2.29}$$

How should we choose ϕ? If we impose $\phi'(0) = \phi'(1) = 0$, then in the one-dimensional case the new integrand (2.27) will vanish at 0 and 1, and so will have a continuous periodic extension. The simplest polynomial choice with this property is

$$\phi(t) = 3t^2 - 2t^3, \quad 0 \leqslant t \leqslant 1, \tag{2.30}$$

for which the derivative is

$$\phi'(t) = 6t(1 - t), \quad 0 \leqslant t \leqslant 1,$$

The graphs of ϕ and ϕ' for this case are shown in Figure 2.10.

In the next figure, Figure 2.11, we show the resulting periodically extended integrand g for this choice of ϕ and a simple choice of the original integrand f. (In fact we took $f(x) = 1 + x$, $0 \leqslant x \leqslant 1$.)

If, as in the last example, f' is well behaved on $(0, 1)$, then it might be sensible to make the periodically extended integrand g smoother than in Figure 2.11. From (2.27), one way to achieve this is to require not only $\phi'(0) = \phi'(1) = 0$, but also $\phi''(0) = \phi''(1) = 0$. The simplest polynomial choice of ϕ with this property is

$$\phi(t) = t^3(10 - 15t + 6t^2) \tag{2.31}$$

corresponding to

$$\phi'(t) = 30t^2(1 - t)^2.$$

Figure 2.12 gives the graphs of ϕ and ϕ' for this case, and Figure 2.13 shows the periodically continued function g, again for the simple choice $f(x) = 1 + x$. Now g is continuously differentiable, as well as periodic.

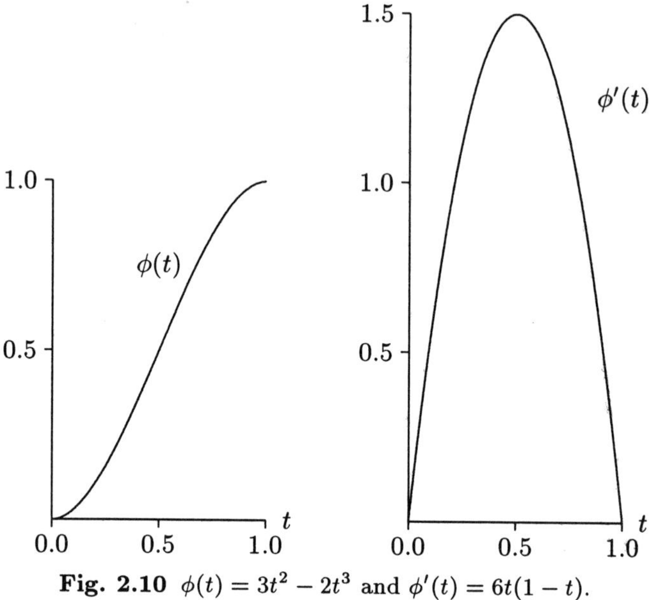

Fig. 2.10 $\phi(t) = 3t^2 - 2t^3$ and $\phi'(t) = 6t(1-t)$.

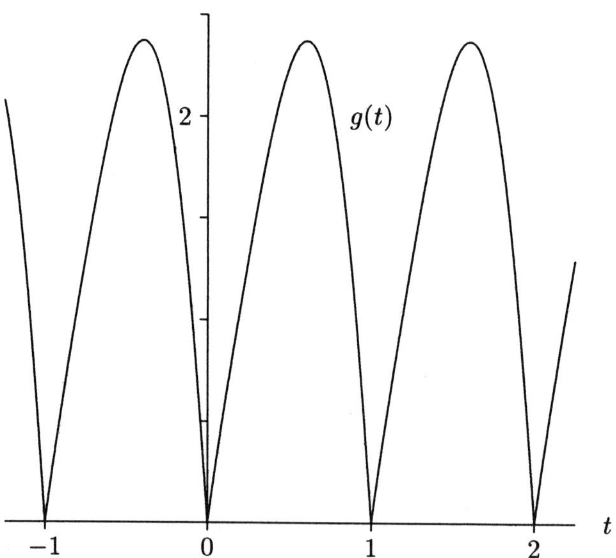

Fig. 2.11 The transformed integrand g for ϕ as in Figure 2.10 and $f(x) = 1 + x$.

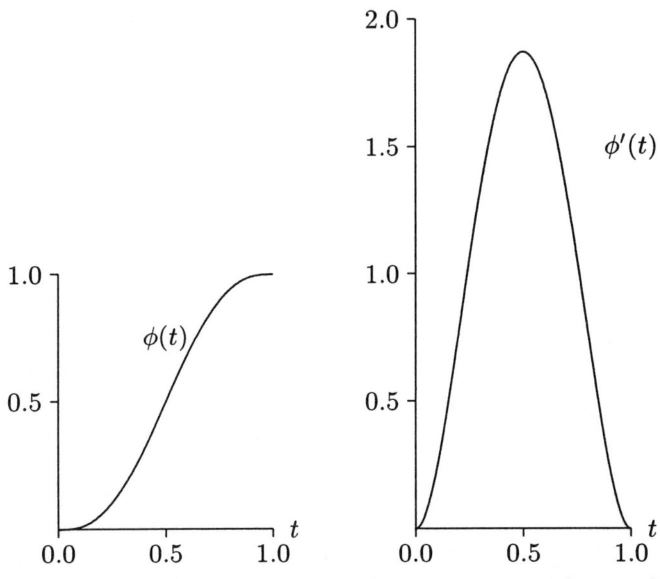

Fig. 2.12 $\phi(t) = t^3(10 - 15t + 6t^2)$ and $\phi'(t) = 30t^2(1-t)^2$.

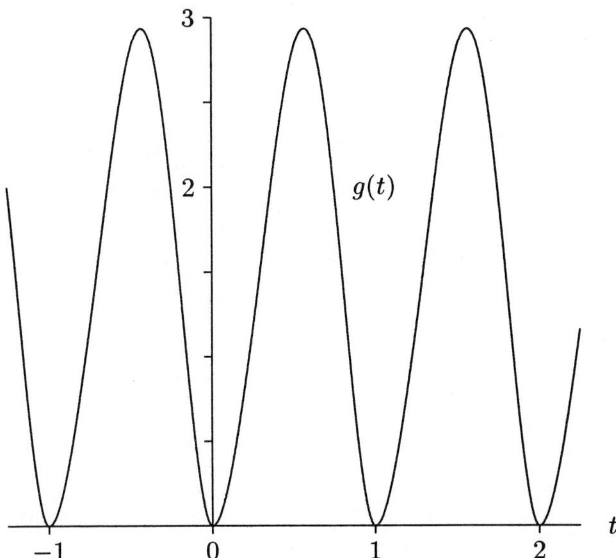

Fig. 2.13 The transformed integrand g for ϕ as in Figure 2.12 and $f(x) = 1 + x$.

For a suitably smooth function f this process can easily be extended as far as desired, by using transformations ϕ of higher and higher degree.

Very recently Sidi (1993) has proposed a different family of transformation functions with a trigonometric rather than polynomial character. The simplest of these, and the only one we consider, is

$$\phi(t) = t - \frac{1}{2\pi} \sin 2\pi t, \qquad (2.32)$$

corresponding to

$$\phi'(t) = 1 - \cos 2\pi t.$$

This transformation, shown in Figure 2.14, is analogous to (2.31) in that $\phi'(t) = O(t^2)$ as $t \to 0$. For us it has an attraction not shared by the traditional transformations: the derivative $\phi'(t)$, which provides an important part of the integrand in (2.29), is a trigonometric polynomial (of degree 1), and trigonometric polynomials are just the functions which lattice rules are designed to handle. Precisely, in the particular case in which f is a constant we see from (2.29) that the lattice rule will give the exact result, provided it integrates exactly all trigonometric polynomials of degree $\leqslant 1$ in each variable. Already we know from Lemma 2.7 that this will be the case provided the dual lattice $L^{\perp}$ has no points $\mathbf{h}$ (other than $\mathbf{0}$) with all components equal to 0 or ± 1—and a lattice rule with such points in its dual lattice would be very bad from every point of view!

The following proposition pushes this result even further.

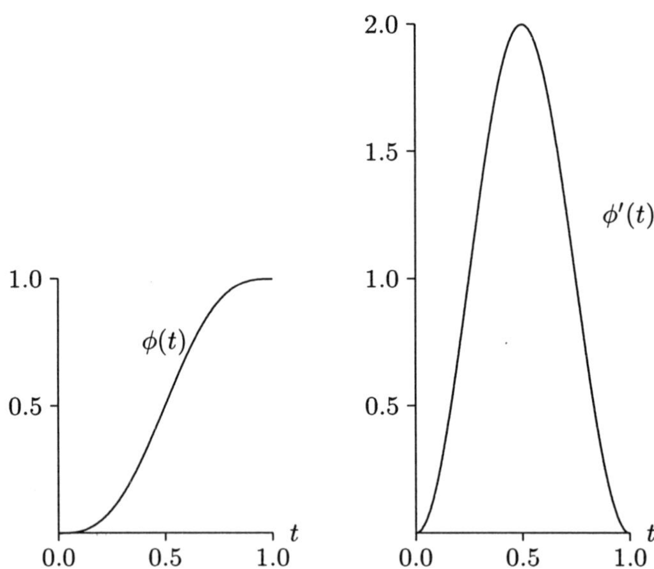

Fig. 2.14 $\phi(t) = t - \sin(2\pi t)/(2\pi)$ and $\phi'(t) = 1 - \cos(2\pi t)$.

Proposition 2.11. *Suppose that a lattice rule Q corresponding to an integration lattice L is applied to the right-hand side of (2.29), where ϕ is given by (2.32). Suppose also that $L^{\perp}$ contains no points (other than $\mathbf{0}$) with all components 0 or ± 1. Then $Qf = If$ for every linear function f.*

Proof The proof for the case $f = $ constant has already been indicated. Now consider the linear function

$$f(\mathbf{x}) = a_1 x_1 + a_2 x_2 + \ldots + a_s x_s + b,$$

which may be rewritten as

$$f(\mathbf{x}) = a_1(x_1 - \frac{1}{2}) + a_2(x_2 - \frac{1}{2}) + \ldots + a_s(x_s - \frac{1}{2}) + c,$$

where

$$c = b + \frac{1}{2}(a_1 + \ldots + a_s).$$

The lattice rule integrates the constant term c exactly, and also gives the exact result 0 for the term $x_1 - \frac{1}{2}$, since if $(x_1, x_2, \ldots, x_s)$ is a point of the integration lattice so is $(1 - x_1, 1 - x_2, \ldots, 1 - x_s)$, and the contribution from these two points cancels. The same argument also holds for the remaining terms, so the result is proved. $\blacksquare$

For completeness we mention two other methods discussed by Zaremba (1972). Restricting ourselves to the one-dimensional integral (2.25), the method of complete symmetrization replaces the given problem by

$$If = \frac{1}{2} \int_0^1 [f(x) + f(1 - x)] \, dx.$$

The generalization to higher dimensions is obvious, but now requires 2^s terms in the sum. This method is usually thought too expensive, especially since it still yields an integrand with no continuity of the first derivatives.

In the Bernoulli polynomial method the integrand is modified by subtraction of a suitable function chosen so that the value of the integral is unchanged, and so that the new integrand has a continuous periodic extension and is as smooth as desired.

Once more it is instructive to demonstrate the one-dimensional case first. Suppose that f has α continuous derivatives on $[0, 1]$, with $\alpha \geqslant 1$. The modified integrand is

$$\psi(x) = f(x) - \sum_{j=1}^{\alpha} \left[f^{(j-1)}(1) - f^{(j-1)}(0) \right] b_j(x), \qquad (2.33)$$

where

$$b_j(x) = \frac{B_j(x)}{j!},$$

and B_j is the Bernoulli polynomial of degree j. The necessary properties of Bernoulli polynomials are stated in Appendix C. Since

$$\int_0^1 b_j(x)\,dx = 0 \text{ for } j \geqslant 1, \tag{2.34}$$

we learn from (2.33) that

$$\int_0^1 \psi(x)\,dx = \int_0^1 f(x)\,dx,$$

so that the value of the integral is unchanged when f is replaced by ψ.

For theoretical purposes, it is convenient to rewrite (2.33), with the aid of the fundamental theorem of calculus, as

$$\psi(x) = f(x) - \sum_{j=1}^{\alpha}\left(\int_0^1 f^{(j)}(x')\,dx'\right) b_j(x). \tag{2.35}$$

The renormalized Bernoulli polynomial b_j has the property, for $j \geqslant 0$ and $k \geqslant 0$,

$$\int_0^1 b_j^{(k)}(x)\,dx = \left\{ \begin{array}{ll} 1, & \text{if } k = j, \\ 0, & \text{otherwise.} \end{array} \right. \tag{2.36}$$

Thus for $k = 1, \ldots, \alpha$ we have

$$\begin{aligned}
\int_0^1 \psi^{(k)}(x)\,dx &= \int_0^1 f^{(k)}(x) - \sum_{j=1}^{\alpha}\left(\int_0^1 f^{(j)}(x')\,dx'\right)\int_0^1 b_j^{(k)}(x)\,dx \\
&= \int_0^1 f^{(k)}(x)\,dx - \int_0^1 f^{(k)}(x')\,dx' = 0,
\end{aligned}$$

or, equivalently,

$$\psi^{(k-1)}(0) = \psi^{(k-1)}(1) \text{ for } 1 \leqslant k \leqslant \alpha.$$

Thus ψ has a continuous periodic extension, with $\alpha - 1$ continuous derivatives.

For example, setting $\alpha = 2$ we have

$$\psi(x) = f(x) - [f(1) - f(0)]\left(x - \frac{1}{2}\right) - [f'(1) - f'(0)]\frac{1}{2}\left(x^2 - x + \frac{1}{6}\right),$$

so that

$$\psi(0) = \psi(1) = \frac{1}{2}\left(f(0) + f(1)\right) + \frac{1}{12}\left(f'(0) - f'(1)\right),$$

$$\psi'(0) = \psi'(1) = f(0) - f(1) + \frac{1}{2}\left(f'(0) + f'(1)\right).$$

A rather nice property of the formula (2.33) is that $\psi \equiv If$ if f is a polynomial of degree $\leqslant \alpha$. This follows easily from (2.35) and (2.36) by expressing f as a linear combination of Bernoulli polynomials.

The formula (2.33) has the obvious difficulty that derivatives of f on the boundary are often not available.

The higher-dimensional generalization of the Bernoulli polynomial method (Korobov 1963, Zaremba 1972) employs a similar construction. Let us assume that all the partial derivatives of f up to $f^{(\alpha,\alpha,\ldots,\alpha)}$ are continuous on C^s, for some $\alpha \geqslant 1$, where we have used the notation

$$\frac{\partial^{j_1 + \cdots + j_s} f}{\partial x_1^{j_1} \cdots \partial x_s^{j_s}} = f^{(j_1,\ldots,j_s)}.$$

(Actually it is sufficient that the partial derivatives are of bounded variation in the sense of Hardy and Krause—see Section 4.2.)

In the two-dimensional case the generalization of (2.33) is

$$
\begin{aligned}
\psi(x_1, x_2) \;=\; & f(x_1, x_2) - \sum_{j=1}^{\alpha} \left[f^{(j-1,0)}(1, x_2) - f^{(j-1,0)}(0, x_2) \right] b_j(x_1) \\[2mm]
& - \sum_{j=1}^{\alpha} \left[f^{(0,j-1)}(x_1, 1) - f^{(0,j-1)}(x_1, 0) \right] b_j(x_2) \\[2mm]
& + \sum_{j_2=1}^{\alpha} \sum_{j_1=1}^{\alpha} \Big[f^{(j_1-1,j_2-1)}(1, 1) - f^{(j_1-1,j_2-1)}(1, 0) \\[2mm]
& \qquad - f^{(j_1-1,j_2-1)}(0, 1) + f^{(j_1-1,j_2-1)}(0, 0) \Big] b_{j_1}(x_1) b_{j_2}(x_2).
\end{aligned}
$$

$$(2.37)$$

Again it follows from (2.34) that $I\psi = If$ so that the value of the integral is unchanged. The expression for ψ in (2.37) may be written as

$$
\begin{aligned}
\psi(x_1, x_2) \;=\; & f(x_1, x_2) - \sum_{j=1}^{\alpha} \left(\int_0^1 f^{(j,0)}(x_1', x_2) \, dx_1' \right) b_j(x_1) \\[2mm]
& - \sum_{j=1}^{\alpha} \left(\int_0^1 f^{(0,j)}(x_1, x_2') \, dx_2' \right) b_j(x_2) \\[2mm]
& + \sum_{j_2=1}^{\alpha} \sum_{j_1=1}^{\alpha} If^{(j_1,j_2)} b_{j_1}(x_1) b_{j_2}(x_2),
\end{aligned}
$$

$$(2.38)$$

which on differentiating k times with respect to x_1, where $1 \leqslant k \leqslant \alpha$, and then integrating with respect to x_1 yields, with the help of (2.36),

$$\int_0^1 \psi^{(k,0)}(x_1,x_2)\,dx_1 = \int_0^1 f^{(k,0)}(x_1,x_2)\,dx_1 - \int_0^1 f^{(k,0)}(x_1',x_2)\,dx_1'$$

$$- \sum_{j=1}^{\alpha} I f^{(k,j)} b_j(x_2) + \sum_{j_2=1}^{\alpha} I f^{(k,j_2)} b_{j_2}(x_2)$$

$$= 0,$$

or

$$\psi^{(k-1,0)}(0,x_2) = \psi^{(k-1,0)}(1,x_2).$$

Thus ψ has a periodic extension with respect to x_1 with $k-1$ continuous derivatives; and, of course, it has a similar property with respect to x_2.

If f is a polynomial of degree $\leqslant \alpha$ in each of x_1 and x_2 separately, then by substitution of f expressed in the form

$$f(x_1,x_2) = \sum_{j_2=0}^{\alpha} \sum_{j_1=0}^{\alpha} a_{j_1 j_2} b_{j_1}(x_1) b_{j_2}(x_2)$$

into (2.38), and use of (2.36), it follows, as in the one-dimensional case, that $\psi \equiv If \ (= a_{00})$.

In higher dimensions the analogous modified integrand can be built up recursively: following Beckers and Haegemans (1992a), we define $\psi_0(\mathbf{x}) := f(\mathbf{x})$ and then determine $\psi_1, \ldots, \psi_s$ by

$$\psi_i(\mathbf{x}) = \psi_{i-1}(\mathbf{x}) - \sum_{j=1}^{\alpha} \left(\left. \frac{\partial^{j-1} \psi_{i-1}}{\partial x_i^{j-1}} \right|_{x_i=1} - \left. \frac{\partial^{j-1} \psi_{i-1}}{\partial x_i^{j-1}} \right|_{x_i=0} \right) b_j(x_i)$$

$$= \psi_{i-1}(\mathbf{x}) - \sum_{j=1}^{\alpha} \left(\int_0^1 \psi_{i-1}^{(0,\ldots,0,j,0,\ldots,0)}(\mathbf{x})\,dx_i' \right) b_j(x_i)$$

for $i = 1, \ldots, s$. It follows by an argument similar to that in the two-dimensional case that $\psi \equiv \psi_s$ has a continuous and appropriately smooth periodic extension with respect to each of $x_1, \ldots, x_s$. Moreover, as before, it follows easily that if f is a polynomial of degree $\leqslant \alpha$ in each of $x_1, \ldots, x_s$ separately, then $\psi \equiv If$.

Even in the two-dimensional case the Bernoulli polynomial method presents difficulties, and the difficulties grow rapidly as s increases. The difficulties lie in the large number of partial derivatives of f which have to be calculated on the boundary of the unit cube C^s. Only in the simplest cases are the derivatives available analytically, and, as Beckers and Haegemans (1992a) have pointed out, even with a symbolic computing facility the problem rapidly becomes difficult or impossible. By the time one has to contemplate numerical approximation to some or all of the derivatives on the boundary, the charm that lies in the formulas has long faded.

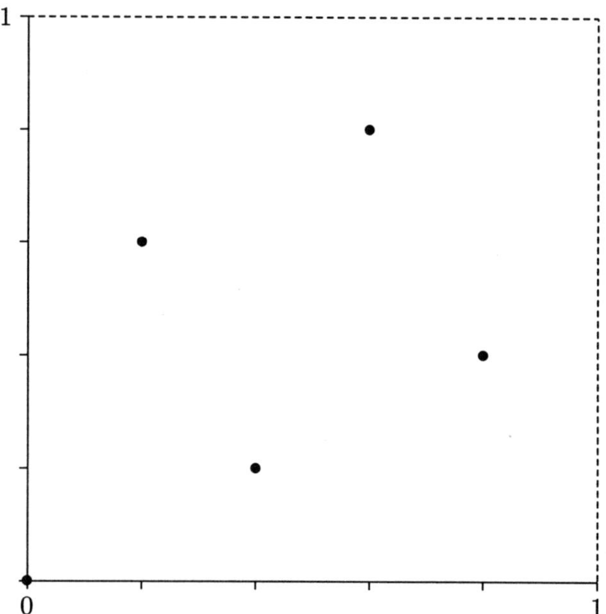

Fig. 2.15 The quadrature points for a two-dimensional lattice rule.

2.13 Geometrically equivalent lattice rules

Sometimes it is useful to think of two lattice rules as being essentially the same if one is carried into the other by one of the symmetry operations of the integration region, the unit cube.

Definition 2.12. *Two lattice rules are* geometrically equivalent *if one can be changed to the other by a relabelling of coordinates, or by replacing the coordinate x_i by $1 - x_i$ (unless $x_i = 0$, in which case it is left unchanged), or by a combination of such transformations.*

Example 2.13. *The lattice rule*

$$Qf = \frac{1}{5}\left[f(0,0) + f\left(\frac{1}{5},\frac{3}{5}\right) + f\left(\frac{2}{5},\frac{1}{5}\right) + f\left(\frac{3}{5},\frac{4}{5}\right) + f\left(\frac{4}{5},\frac{2}{5}\right) \right]$$

is geometrically equivalent to the rule given by (2.13), since it may be obtained from (2.13) by replacing x_1 by $1 - x_1$, or, alternatively, by interchanging the first and second coordinates. The quadrature points for this rule are given in Figure 2.15. The reader will agree, we think, that this rule does not differ in a very interesting way from that shown in Figure 2.1.

3

Lattice rules as multiple sums

3.1 Introduction

In the preceding chapter we have seen that a lattice rule Q is an equal-weight quadrature rule in which the quadrature points are all the points belonging to some integration lattice L that lie in U^s. However, that rather abstract definition does not tell us how lattice rules may be written down, or implemented in a computer program. Our main task in this chapter is to obtain a mathematical classification that is at the same time useful for computation.

The first point to notice is that a given lattice rule can be expressed in many different ways. This is demonstrated by the following example.

Example 3.1. *Consider the lattice rule of the form* (2.5) *with $N = 5$ and* $\mathbf{z} = (1, 2)$,

$$Qf = \frac{1}{5} \sum_{j=0}^{4} f\left(\left\{\frac{j}{5}(1, 2)\right\}\right),\tag{3.1}$$

where the braces around a vector denote, as in Definition 2.9, that the fractional part of each component is to be taken. The five quadrature points of this two-dimensional lattice rule, namely $(0, 0)$, $\left(\frac{1}{5}, \frac{2}{5}\right)$, $\left(\frac{2}{5}, \frac{4}{5}\right)$, $\left(\frac{3}{5}, \frac{1}{5}\right)$, $\left(\frac{4}{5}, \frac{3}{5}\right)$, were shown previously in Figure 2.1. However, the same five quadrature points, and hence the same lattice rule, are also given by the formula

$$Qf = \frac{1}{5} \sum_{j=0}^{4} f\left(\left\{\frac{j}{5}(q, 2q)\right\}\right),\tag{3.2}$$

with q taking any of the values 2, 3, or 4. (To see this, just write out the quadrature points: the same ones occur in each case, but in a different order.) Moreover, the lattice rule (3.1) with five quadrature points can be expressed in a repetitive form in which each quadrature point occurs more than once. For instance, the lattice rules

$$Qf = \frac{1}{10} \sum_{j=0}^{9} f\left(\left\{\frac{j}{10}(2, 4)\right\}\right)$$

and

$$Qf = \frac{1}{25} \sum_{j_2=0}^{4} \sum_{j_1=0}^{4} f\left(\left\{\frac{j_1}{5}(1,2) + \frac{j_2}{5}(3,1)\right\}\right)$$

are easily seen to be equivalent to (3.1), with each quadrature point occurring twice in the first expression and five times in the second.

In the next section, the most important one in this chapter, we look at the structure and classification of lattice rules. This is done by giving a canonical representation of lattice rules based on multiple sums, from the work of Sloan and Lyness (1989). This canonical form provides a classification of lattice rules based on the concepts of 'rank' and 'invariants'. In simple terms, the rank of a lattice rule is the minimum number of sums required to write it down. For example, the lattice rule in Example 3.1 is a rule of rank 1: the formulas (3.1) and (3.2) both express the rules in the canonical form for rules of rank 1. (There is only one sum!) Elementary group theory is used in obtaining the canonical form in the general case; for readers not familiar with this topic, a simple introduction to group theory is included.

Section 3.3 introduces the notion of the 'direct sum' of lattice rules, one approach to building a lattice rule from rules of lower order.

Section 3.4 looks at the 'projections' of a lattice rule, that is, at lower-dimensional rules obtained by neglecting one or more components. It turns out that the rank and invariants of a projection are constrained in a simple way by those of the original rule.

If the rank of a lattice rule is r, then the canonical form requires r sums and r integer vectors $z_1, \ldots, z_r$. In general, for a given lattice rule these integer vectors are not uniquely determined (as may be seen in the equivalent rank-1 rules given by (3.1) and (3.2)). However, the results of Sloan and Lyness (1990) show that for one particular class of lattice rules the integer vectors $z_1, \ldots, z_r$ may be chosen uniquely. These uniqueness results are looked at in Section 3.5. The particular class of rules for which this uniqueness property holds are the 'projection-regular' lattice rules. Put simply, a projection-regular rule is one in which for $1 \leqslant k \leqslant s-1$ the lattice rules obtained by dropping the last k components have as many quadrature points as possible, given full knowledge of the rank and invariants of the original rule.

Finally, for the particular case of rank-2 lattice rules, a different unique representation obtained by Lyness and Sloan (1989) is considered in Section 3.6. Unlike the results of Section 3.5, Section 3.6 is concerned with the unique representation obtained when a rank-2 rule is written with three sums, one more than the double sum required for the canonical form. This 'tricycle' form has certain advantages for computer searches of rank-2 rules.

3.2   Canonical form for lattice rules

A key result is that every lattice rule can be written in the 'canonical form' expressed by the following theorem.

Theorem 3.2. *Let Q be any s-dimensional lattice rule of order $N \geqslant 2$. Then there exists a uniquely determined integer r (the 'rank') with $1 \leqslant r \leqslant s$ and uniquely determined integers $n_1, \ldots, n_r > 1$ (the 'invariants') satisfying*

$$n_{k+1} \text{ divides } n_k, \quad 1 \leqslant k \leqslant r - 1,$$

such that Q can be expressed in the form

$$Qf = \frac{1}{N} \sum_{j_r=0}^{n_r-1} \cdots \sum_{j_1=0}^{n_1-1} f\left(\left\{\frac{j_1}{n_1}\mathbf{z}_1 + \cdots + \frac{j_r}{n_r}\mathbf{z}_r\right\}\right), \qquad (3.3)$$

where $\mathbf{z}_1, \ldots, \mathbf{z}_r$ are linearly independent integer vectors, and the order of the rule is

$$N = n_1 n_2 \cdots n_r.$$

The linear independence of $\mathbf{z}_1, \ldots, \mathbf{z}_r$ asserted by the theorem is ordinary linear independence with respect to the rational numbers.

Theorem 3.2 excludes the simplest lattice rule of all, the trivial one-point rule

$$Qf = f(\mathbf{0}).$$

It is convenient to define the rank of this one-point rule to be 0.

The proof of Theorem 3.2 is based on results from group theory, found in standard texts such as Ledermann (1964). Since the reader may not be familiar with group theory, we first give a simple introduction to this area of mathematics, with examples illustrating the application to lattice rules. Some readers may prefer to pass directly to the proof of Theorem 3.2, which follows Theorem 3.22.

We start with the definition of a group.

Definition 3.3. *A* group *$\mathcal{G}$ is a set together with a single-valued binary operation $\circ$ such that:*

(i) *The set is closed under the operation $\circ$.*

(ii) *The operation $\circ$ is associative.*

(iii) *There exists an identity element $e \in \mathcal{G}$ such that $e \circ g = g = g \circ e$ for all $g \in \mathcal{G}$.*

(iv) *There exists, for each $g \in \mathcal{G}$, an inverse element $g^{-1} \in \mathcal{G}$ such that $g \circ g^{-1} = e = g^{-1} \circ g$.*

The group $\mathcal{G}$ is said to be abelian *if the operation $\circ$ is also commutative; that is, for all $g, g' \in \mathcal{G}$, $g \circ g' = g' \circ g$. A* finite group *is one with a finite number of elements; otherwise the group is an* infinite group. *If a group is finite, the number of its elements is called the* order *of the group.*

All the groups considered here will be finite and abelian, and have order $\geqslant 2$.

The following result shows the relevance of finite abelian groups to the theory of lattice rules. (For the definition of lattice rules and integration lattices, see Definitions 2.3 and 2.2.)

Theorem 3.4. *The elements* $\mathbf{x}_0, \ldots, \mathbf{x}_{N-1}$ *belonging to* $\mathcal{A}(Q)$, *the set of quadrature points of a lattice rule* Q *of order* N, *form an abelian group of order* N *under the operation*

$$\mathbf{x}_j \circ \mathbf{x}_k = \{\mathbf{x}_j + \mathbf{x}_k\}, \quad 0 \leqslant j, k \leqslant N - 1. \tag{3.4}$$

Proof Suppose L is the integration lattice corresponding to Q. Because L is a lattice, the sum and difference of any two points in L is also in L. Hence if $\mathbf{x}_j, \mathbf{x}_k \in \mathcal{A}(Q) \subset L$, then $\mathbf{x}_j + \mathbf{x}_k \in L$. Since $\{\mathbf{x}_j + \mathbf{x}_k\} \in U^s$ is obtained from $\mathbf{x}_j + \mathbf{x}_k$ by subtraction of an appropriate integer vector, and since every integer vector belongs to L, we conclude that $\{\mathbf{x}_j + \mathbf{x}_k\} \in L$, and hence

$$\{\mathbf{x}_j + \mathbf{x}_k\} \in L \cap U^s = \mathcal{A}(Q).$$

Thus $\mathcal{A}(Q)$ is closed under the group operation (3.4). Also, for elements $\mathbf{x}_j, \mathbf{x}_k, \mathbf{x}_\ell \in \mathcal{A}(Q)$, we have

$$\{\mathbf{x}_j + \{\mathbf{x}_k + \mathbf{x}_\ell\}\} = \{\mathbf{x}_j + \mathbf{x}_k + \mathbf{x}_\ell\} = \{\{\mathbf{x}_j + \mathbf{x}_k\} + \mathbf{x}_\ell\},$$

and so the group operation is associative. Obviously, it is also commutative. Moreover, it is clear that $\mathbf{0}$ is an identity element. Finally, every element $\mathbf{x}_j \in \mathcal{A}(Q)$ has $\{-\mathbf{x}_j\}$ as an inverse element. ∎

In words, the group operation for the set of quadrature points is addition 'modulo 1', meaning that two vectors are treated as identical if they differ by an integer vector.

Example 3.5. *For the lattice rule given in Example* 3.1 *the set of quadrature points is*

$$\mathcal{A}(Q) = \left\{ (0,0), \left(\frac{1}{5}, \frac{2}{5}\right), \left(\frac{2}{5}, \frac{4}{5}\right), \left(\frac{3}{5}, \frac{1}{5}\right), \left(\frac{4}{5}, \frac{3}{5}\right) \right\}.$$

Regarded as a group, the identity element is $(0,0)$. *The group operation, of addition modulo* 1, *is exemplified by*

$$\left(\frac{2}{5}, \frac{4}{5}\right) \circ \left(\frac{4}{5}, \frac{3}{5}\right) = \left(\frac{1}{5}, \frac{2}{5}\right).$$

The inverse of $\left(\frac{2}{5}, \frac{4}{5}\right)$ *in the group* $\mathcal{A}(Q)$ *is* $\left\{\left(-\frac{2}{5}, -\frac{4}{5}\right)\right\} = \left(\frac{3}{5}, \frac{1}{5}\right)$, *since*

$$\left(\frac{2}{5}, \frac{4}{5}\right) \circ \left(\frac{3}{5}, \frac{1}{5}\right) = (0,0).$$

We shall need a few more definitions. The first is concerned with the notion of a subgroup.

Definition 3.6. *A subset $\mathcal{G}'$ of a group $\mathcal{G}$ is a subgroup of $\mathcal{G}$ if $\mathcal{G}'$ forms a group under the group operation of $\mathcal{G}$.*

In Example 3.5, the only subgroups of $\mathcal{A}(Q)$ are $\mathcal{A}(Q)$ itself, and the 'trivial subgroup', which contains only the identity element $(0,0)$. Such subgroups are not very interesting. The next example is more interesting, in that it contains 'proper' subgroups, that is, subgroups other than $\mathcal{A}(Q)$ and the trivial subgroup.

Example 3.7. *It is easily verified that*

$$\mathcal{A}(Q) = \left\{ (0,0), \left(0, \tfrac{1}{2}\right), \left(\tfrac{1}{2}, 0\right), \left(\tfrac{1}{2}, \tfrac{1}{2}\right) \right\}$$

is a group under the group operation (3.4). The subgroups of $\mathcal{A}(Q)$ are:

$$\mathcal{A}(Q), \quad \left\{ (0,0), \left(0, \tfrac{1}{2}\right) \right\}, \quad \left\{ (0,0), \left(\tfrac{1}{2}, 0\right) \right\}, \quad \left\{ (0,0), \left(\tfrac{1}{2}, \tfrac{1}{2}\right) \right\}, \quad \{(0,0)\}.$$

Sometimes a group can be expressed in a nice way in terms of some of its subgroups.

Definition 3.8. *An abelian group $\mathcal{G}$ with group operation $\circ$ is said to be the direct sum of the subgroups $\mathcal{G}_1, \ldots, \mathcal{G}_t$ if each $g \in \mathcal{G}$ can be expressed uniquely in the form $g = g_1 \circ \cdots \circ g_t$, where $g_j \in \mathcal{G}_j$, $1 \leqslant j \leqslant t$. In this case, we write*

$$\mathcal{G} = \mathcal{G}_1 \oplus \cdots \oplus \mathcal{G}_t.$$

A similar definition also holds for nonabelian groups, but in that case one usually prefers to speak of a 'direct product' rather than a 'direct sum'.

Example 3.9. *For the preceding Example 3.7 the reader will easily verify that*

$$\mathcal{A}(Q) = \left\{ (0,0), \left(0, \tfrac{1}{2}\right) \right\} \oplus \left\{ (0,0), \left(\tfrac{1}{2}, 0\right) \right\}.$$

For example, $\left(0, \tfrac{1}{2}\right) = \left(0, \tfrac{1}{2}\right) \circ (0,0)$ and $\left(\tfrac{1}{2}, \tfrac{1}{2}\right) = \left(0, \tfrac{1}{2}\right) \circ \left(\tfrac{1}{2}, 0\right)$. Moreover, the four elements found by adding an element of the first subgroup to an element of the second are just the four elements of $\mathcal{A}(Q)$. Interestingly, it is equally easy to verify that

$$\mathcal{A}(Q) = \left\{ (0,0), \left(\tfrac{1}{2}, 0\right) \right\} \oplus \left\{ (0,0), \left(\tfrac{1}{2}, \tfrac{1}{2}\right) \right\}.$$

This illustrates an important point: that the component subgroups in a direct sum decomposition are in general not uniquely determined.

Two groups with different elements and different group operations may still have the same essential structure. This leads to the concept of the 'isomorphism' of groups.

Definition 3.10. *Two groups $\mathcal{G}$ and $\mathcal{G}'$ with group operations $\circ$ and $\odot$, respectively, are* isomorphic *if there is a one-to-one correspondence between the elements of $\mathcal{G}$ and those of $\mathcal{G}'$, such that if $g \in \mathcal{G}$ corresponds to $g' \in \mathcal{G}'$, and $h \in \mathcal{G}$ corresponds to $h' \in \mathcal{G}'$, then $g \circ h \in \mathcal{G}$ corresponds to $g' \odot h' \in \mathcal{G}'$. One says that there is an* isomorphism *from $\mathcal{G}$ to $\mathcal{G}'$.*

In words, two groups are isomorphic if they have the same 'multiplication table' (or, as might be preferred for abelian groups, the same 'addition table') after an appropriate relabelling.

Example 3.11. *The groups*

$$\left\{(0,0), \left(\frac{1}{5},\frac{2}{5}\right), \left(\frac{2}{5},\frac{4}{5}\right), \left(\frac{3}{5},\frac{1}{5}\right), \left(\frac{4}{5},\frac{3}{5}\right)\right\},$$

$$\left\{(0,0), \left(\frac{1}{5},\frac{3}{5}\right), \left(\frac{2}{5},\frac{1}{5}\right), \left(\frac{3}{5},\frac{4}{5}\right), \left(\frac{4}{5},\frac{2}{5}\right)\right\},$$

each with the group operation (3.4), are isomorphic. The one-to-one correspondence may be taken, for example, to be between the kth element in the first group and the kth element in the second group. Both groups are also isomorphic to the group formed by the five fifth roots of unity under the operation of ordinary multiplication of complex numbers. Indeed, every group of order 5 is isomorphic to every other group of the same order. This property extends to groups of any prime order. To a mathematician, groups of prime order are rather dull, because there is really only one abstract group of each prime order.

We now look at the 'powers' of the elements of a group.

Definition 3.12. *Suppose $\mathcal{G}$ is a group with group operation $\circ$. For $g \in \mathcal{G}$ and any positive integer q, define*

$$g^q = \underbrace{g \circ g \circ \cdots \circ g}_{q \text{ times}},$$

with $g^0 = e$, where e is the identity element. An element $g \in \mathcal{G}$ is said to be of order n *if n is the least positive integer for which $g^n = e$.*

As an example, with the group operation given by (3.4), it is easily seen that $\mathbf{x}^q = \{q\mathbf{x}\}$. It is also possible to define negative powers: for any positive integer q, $g^{-q} \in \mathcal{G}$ is the element for which $g^{-q}g^q = e$. That is, g^{-q} is the inverse of g^q.

Example 3.13. *In Example 3.5 the point $(\frac{1}{5}, \frac{2}{5})$ is of order 5, as is every point other than $(0,0)$, which of course is of order 1. In Example 3.7 the point $(0, \frac{1}{2})$ is of order 2. It is an elementary theorem of group theory that the order of every element divides the order of the group.*

An important special case of an abelian group is a 'cyclic' group.

Definition 3.14. *A group $\mathcal{G}$ of order N is said to be* cyclic *if there exists an element $u \in \mathcal{G}$ (a* generator *of the group) such that every element $g \in \mathcal{G}$ may be written as $g = u^q$ for some nonnegative integer q.*

Note that a generator of a cyclic group of order N must itself be of order N, that is, N must be the smallest number q such that $u^q = e$, with e the identity element. Less obvious but true is that in any group of order N *every* element of order N is a generator, and the existence of even one element of order N makes the group cyclic. (One simply observes that in this situation each power u^q with $1 \leqslant q \leqslant N$ must be different.)

Example 3.15. *We have already seen that the quadrature points of the lattice rule (3.1) form a group under the operation given by (3.4). This group is cyclic: the generator may be taken to be $(\frac{1}{5}, \frac{2}{5})$. On the other hand, the group in Example 3.7 is not cyclic, because the group has no element of order 4, and hence has no element that could serve as generator of the group.*

We remark that if the element u is a generator of a cyclic group of order N, then u^q is also a generator provided q and N are relatively prime. This explains why the vectors $(\frac{q}{5}, \frac{2q}{5})$ with $q = 2, 3$, or 4 given in (3.2) generate the quadrature points for the lattice rule (3.1).

From Theorem 3.4 we see that the structure of a lattice rule is, in essence, the structure of $\mathcal{A}(Q)$. In particular, various decompositions of $\mathcal{A}(Q)$ into the direct sum of cyclic groups will play a crucial role. We now look at the required decomposition theory of abelian groups.

Suppose $\mathcal{G}$ is an abelian group of order N. If N has the prime factorization $N = p_1^{\gamma_1} p_2^{\gamma_2} \cdots p_q^{\gamma_q}$, where $p_1, \ldots, p_q$ are distinct primes, then it is known that $\mathcal{G}$ may be expressed as the direct sum of q (abelian) subgroups,

$$\mathcal{G} = \mathcal{S}_1 \oplus \cdots \oplus \mathcal{S}_q, \tag{3.5}$$

where $\mathcal{S}_k$ is a group of order $p_k^{\gamma_k}$. The group $\mathcal{S}_k$ consists of the elements of $\mathcal{G}$ whose order is a power of p_k and is known as the Sylow p_k subgroup of $\mathcal{G}$. Note that the group $\mathcal{A}(Q)$ in Example 3.7 is itself a Sylow 2 subgroup, and that for this particular example there are no other Sylow subgroups.

Sylow subgroups may or may not be cyclic—remember that the group in Example 3.7 is not cyclic! Sylow subgroups which are not cyclic may be decomposed further: any abelian group $\mathcal{S}$ of order p^γ, where p is prime, may be expressed as the sum of *cyclic* subgroups,

$$\mathcal{S} = \mathcal{E}_1 \oplus \cdots \oplus \mathcal{E}_{q'}, \tag{3.6}$$

where $\mathcal{E}_j$ has order p^{δ_j} with $\sum_{j=1}^{q'} \delta_j = \gamma$. This direct sum decomposition of prime-power abelian groups is unique up to an isomorphism. This is illustrated nicely by Example 3.9: there we see the group from Example 3.7 expressed as the direct sum of cyclic groups in two different ways. Thus the decomposition is not unique; but it *is* unique up to an isomorphism. It is useful to note that any group of prime order p is already cyclic, and also that it is not possible to decompose a cyclic group of order p^γ any further.

Taking (3.5) and (3.6) together, we find that any abelian group $\mathcal{G}$ of order N may be expressed as the direct sum of *cyclic* subgroups of prime-power order,

$$\mathcal{G} = \mathcal{C}_1 \oplus \cdots \oplus \mathcal{C}_t, \tag{3.7}$$

where the cyclic group $\mathcal{C}_j$ has order $p_j^{\delta_j}$ for some prime p_j and positive integer δ_j and $N = p_1^{\delta_1} p_2^{\delta_2} \cdots p_t^{\delta_t}$. Note that in (3.7) there is no need for the p_j or δ_j to be distinct. By the construction, the number t is uniquely determined, and so, too, up to an isomorphism, are the cyclic groups $\mathcal{C}_1, \ldots, \mathcal{C}_t$ in (3.7).

Example 3.16. *Consider the group of order $12 = 3 \cdot 2^2$ given by*

$$\mathcal{G} = \left\{ (0,0), \left(\frac{1}{6}, \frac{1}{3}\right), \left(\frac{1}{3}, \frac{2}{3}\right), \left(\frac{1}{2}, 0\right), \left(\frac{2}{3}, \frac{1}{3}\right), \left(\frac{5}{6}, \frac{2}{3}\right), \right.$$
$$\left. \left(\frac{1}{2}, \frac{1}{2}\right), \left(\frac{2}{3}, \frac{5}{6}\right), \left(\frac{5}{6}, \frac{1}{6}\right), \left(0, \frac{1}{2}\right), \left(\frac{1}{6}, \frac{5}{6}\right), \left(\frac{1}{3}, \frac{1}{6}\right) \right\}.$$

This group is in fact the set of quadrature points for the lattice rule

$$Qf = \frac{1}{12} \sum_{j_2=0}^{1} \sum_{j_1=0}^{5} f\left(\left\{ \frac{j_1}{6}(1,2) + \frac{j_2}{2}(1,1) \right\}\right). \tag{3.8}$$

The Sylow 3 subgroup is given by

$$\mathcal{S}_1 = \left\{ (0,0), \left(\frac{1}{3}, \frac{2}{3}\right), \left(\frac{2}{3}, \frac{1}{3}\right) \right\} = \mathcal{C}_1,$$

which is already a cyclic group and so cannot be decomposed any further. The Sylow 2 subgroup of $\mathcal{G}$ is

$$\mathcal{S}_2 = \left\{ (0,0), \left(\frac{1}{2}, 0\right), \left(\frac{1}{2}, \frac{1}{2}\right), \left(0, \frac{1}{2}\right) \right\}.$$

As we saw in Example 3.9, this Sylow 2 subgroup may be decomposed in more than one way; one such way is

$$\mathcal{S}_2 = \left\{ (0,0), \left(0, \tfrac{1}{2}\right) \right\} \oplus \left\{ (0,0), \left(\tfrac{1}{2}, 0\right) \right\} = \mathcal{C}_2 \oplus \mathcal{C}_3.$$

Thus a complete decomposition of the form (3.7) is given by

$$\mathcal{G} = \mathcal{C}_1 \oplus \mathcal{C}_2 \oplus \mathcal{C}_3. \tag{3.9}$$

The direct sum decomposition (3.7) into prime-power cyclic groups may seem rather abstract, yet it yields an immediate result for lattice rules. Suppose that the group $\mathcal{G}$ in (3.7) is $\mathcal{A}(Q)$, the set of quadrature points for some lattice rule Q. Then since $\mathcal{C}_j$ is a cyclic group of order $n_j = p_j^{\delta_j}$ for some prime p_j, by definition there exists an s-vector $\mathbf{c}_j$ such that $\{n_j \mathbf{c}_j\} = \mathbf{0}$. In other words, if $\mathbf{z}_j := n_j \mathbf{c}_j$, then $\mathbf{z}_j \in \mathbb{Z}^s$. Thus the generator $\mathbf{c}_j$ of the cyclic subgroup $\mathcal{C}_j$ can be written in the form $\mathbf{z}_j / n_j$. Carrying this out for each value of j from 1 to t, we obtain the following result.

Theorem 3.17. *Any lattice rule Q of order $N \geqslant 2$ may be written as*

$$Qf = \frac{1}{N} \sum_{j_t=0}^{n_t-1} \cdots \sum_{j_1=0}^{n_1-1} f\left(\left\{\frac{j_1}{n_1}\mathbf{z}_1 + \cdots + \frac{j_t}{n_t}\mathbf{z}_t\right\}\right), \tag{3.10}$$

where $N = n_1 n_2 \cdots n_t$ and n_j is the power of a prime. (The primes and prime powers need not be distinct.)

The form of the lattice rule given in (3.10) is not necessarily the most convenient in practice. The value of t may be large, and there is no requirement that the integer vectors $\mathbf{z}_1, \ldots, \mathbf{z}_t$ be linearly independent.

To obtain the canonical form of the lattice rule given in Theorem 3.2, we need to recombine some of the cyclic groups in the decomposition (3.7). Through this recombination, we shall obtain a decomposition involving a smaller number of cyclic groups, whose orders are not necessarily prime powers. Our target is Theorem 3.22 below, which is the fundamental structure theorem for finite abelian groups.

The following well-known lemma is a useful tool in the recombination process.

Lemma 3.18. *Let $\mathcal{C}$ and $\mathcal{C}'$ be cyclic subgroups of orders p and q, respectively, of an abelian group $\mathcal{G}$, having only the zero element in common. Then*

$$\mathcal{D} = \mathcal{C} \oplus \mathcal{C}'$$

is a cyclic group of order pq if p and q are relatively prime. It is not a cyclic group when p and q have a nontrivial common factor.

The lemma follows from the following simple argument. Let $\mathbf{c}$ and $\mathbf{c}'$ be generators of C and C'. Then it is easily verified that $\mathbf{c} \circ \mathbf{c}'$ is of order pq/h, where h is the greatest common divisor of p and q. If p and q are relatively prime, then $pq/h = pq$, the order of $\mathcal{D}$. On the other hand, if $h > 1$ then neither $\mathbf{c} \circ \mathbf{c}'$ nor any other element of $\mathcal{D}$ is of order pq.

Example 3.19. *We may easily verify that the six-point lattice rule*

$$\frac{1}{6} \sum_{j_2=0}^{2} \sum_{j_1=0}^{1} f\left(\left\{\frac{j_1}{2}(1,0) + \frac{j_2}{3}(1,1)\right\}\right)$$

is equivalent to

$$\frac{1}{6} \sum_{j=0}^{5} f\left(\left\{\frac{j}{6}(1,4)\right\}\right).$$

In the first form $\mathcal{A}(Q)$ is in effect expressed as the direct sum of two cyclic groups of relatively prime orders 2 and 3. In the second it is expressed as a single cyclic group of order 6.

When N is a prime power, the decomposition (3.7) represents the unique (up to an isomorphism) decomposition of $\mathcal{G}$ into the direct sum of cyclic subgroups. In all other cases the decomposition into cyclic groups is not unique. In particular, at least two of the cyclic groups in (3.7) have relatively prime orders, and so, from Lemma 3.18, may be recombined into a cyclic group of higher order. Starting from (3.7), we may successively combine cyclic groups of relatively prime orders, a pair at a time, until, at some stage, further recombination is not possible. Thus we end up with a decomposition of $\mathcal{G}$ of the form

$$\mathcal{G} = \mathcal{D}_1 \oplus \cdots \oplus \mathcal{D}_{\tilde{t}}, \tag{3.11}$$

where $\tilde{t} \leqslant t$. Generally, this recombination process is not unique, but it is clear that there must be a recombination which produces a minimum value for $\tilde{t}$. This leads to the concept of 'rank'.

Definition 3.20. *The* rank *of the finite abelian group $\mathcal{G}$ of order $\geqslant 2$ is the minimum number of cyclic subgroups into which $\mathcal{G}$ may be decomposed.*

Thus if r is the rank of the group $\mathcal{G}$, there exists no decomposition of the form (3.11) with $\tilde{t} < r$.

Here is one convenient way of organizing the recombination process. Let p_1 denote the smallest prime that occurs in the factorization of N and place in a horizontal row all of the cyclic subgroups in (3.7) whose order is a power of p_1. Arrange these cyclic groups so that their orders are nonincreasing from left to right. For the next smallest prime p_2 occurring in the factorization of N repeat the process, with the new row placed so that the left-hand members, that is, those with largest order, of the two rows are aligned. Repeat the process until all the primes in N, and hence all the cyclic groups, are exhausted. Then $\mathcal{D}_1$ can be taken to be the direct sum of all the cyclic groups in the left-most *column*, $\mathcal{D}_2$ the direct sum of all those in the next column, and so on, until there are no cyclic groups left.

That construction makes it clear that

$$r = \max_p \mu(p),$$

where $\mu(p)$ is the number of cyclic subgroups in (3.7) whose order is a power of p, and the maximum is taken over all the primes which occur in the factorization of N.

In the above construction it is clear that the order of $\mathcal{D}_2$ divides that of $\mathcal{D}_1$, the order of $\mathcal{D}_3$ divides that of $\mathcal{D}_2$, and so on. (To see this, think of each prime factor of N separately: for we arranged the ordering in each row so that this property already holds for each prime separately.)

Example 3.21. *For the group given in Example* 3.16, *a recombination of* (3.9) *in the manner just described yields*

$$\mathcal{G} = \mathcal{D}_1 \oplus \mathcal{D}_2,$$

where

$$\mathcal{D}_1 = \mathcal{C}_2 \oplus \mathcal{C}_1$$

is the cyclic group of order 6 *generated by* $\left(\frac{1}{3}, \frac{1}{6}\right)$, *and*

$$\mathcal{D}_2 = \mathcal{C}_3 = \left\{ (0,0), \left(\frac{1}{2}, 0\right) \right\}.$$

Thus the rank of the group $\mathcal{G}$ *is* 2. *The orders of* $\mathcal{D}_1$ *and* $\mathcal{D}_2$ *(the 'invariants' of the group) are* 6 *and* 2.

The construction given above leads to a well-known structure theorem for finite abelian groups (for instance, see Ledermann (1964, Section 45, Theorem 4)).

Theorem 3.22. *A finite abelian group* $\mathcal{G}$ *of order* ≥ 2 *may be expressed (uniquely up to an isomorphism) in the form*

$$\mathcal{G} = \mathcal{D}_1 \oplus \cdots \oplus \mathcal{D}_r,$$

where $\mathcal{D}_k$ *is a cyclic group of order* $n_k > 1$, *and*

$$n_{k+1} \text{ divides } n_k, \quad 1 \leqslant k \leqslant r - 1.$$

The numbers r *(the 'rank') and* $n_1, \ldots, n_r$ *(the 'invariants') are uniquely determined.*

Now we return to the proof of Theorem 3.2.

Proof of Theorem 3.2 We apply Theorem 3.22 to the group $\mathcal{A}(Q)$ in Theorem 3.4, and use the fact that, for $1 \leqslant k \leqslant r$, a generator of $\mathcal{D}_k$ (which by definition is a cyclic group of order n_k) can be written in the form $\mathbf{z}_k/n_k$, where $\mathbf{z}_k$ is an integer vector. Then we see that Theorem 3.2 follows immediately, except for the result that the integer vectors $\mathbf{z}_1, \ldots, \mathbf{z}_r$ are linearly independent over the rational numbers.

We now prove that this is indeed the case by establishing an equivalent property, namely that the vectors $\mathbf{z}_k/n_k$, $1 \leqslant k \leqslant r$, are linearly independent. To do this, assume the contrary. Then there exist rational numbers $\lambda_1, \ldots, \lambda_r$, not all zero, such that

$$\frac{\lambda_1}{n_1}\mathbf{z}_1 + \cdots + \frac{\lambda_r}{n_r}\mathbf{z}_r = \mathbf{0}.$$

Without loss of generality we may assume that $\lambda_1, \ldots, \lambda_r$ are integers having no nontrivial common factor. The assumption implies that

$$\left\{\frac{\lambda_1}{n_1}\mathbf{z}_1 + \cdots + \frac{\lambda_r}{n_r}\mathbf{z}_r\right\} = \mathbf{0},$$

and in turn

$$\left\{\frac{\lambda_1 \bmod n_1}{n_1}\mathbf{z}_1 + \cdots + \frac{\lambda_r \bmod n_r}{n_r}\mathbf{z}_r\right\} = \mathbf{0},$$

where $\lambda_k \bmod n_k$ denotes the least residue of λ_k modulo n_k.

The quadrature point $\mathbf{0}$ must occur only once in the rule (3.3), because this rule is a restatement of a direct sum representation of $\mathcal{A}(Q)$. Thus we conclude that $\lambda_k \bmod n_k = 0$, $1 \leqslant k \leqslant r$, since otherwise there would exist nontrivial values of $j_1, \ldots, j_r$ in the sum (3.3) which would make the argument of f equal to $\mathbf{0}$. Equivalently,

$$\lambda_k \text{ is a multiple of } n_k, \quad 1 \leqslant k \leqslant r. \tag{3.12}$$

Since n_{k+1} divides n_k for $1 \leqslant k \leqslant r-1$, the invariants have n_r as a common factor. From (3.12), this means that n_r is a common factor of $\lambda_1, \ldots, \lambda_r$. This contradicts the assumption that the $\lambda_1, \ldots, \lambda_r$ have no nontrivial common factor, and so completes the proof of linear independence. Of course, the property that $r \leqslant s$ is a consequence of the linear independence of the vectors $\mathbf{z}_1, \ldots, \mathbf{z}_r$. $\blacksquare$

3.3 Direct sums of lattice rules

A useful by-product of the preceding section is that it gives us a way of building lattice rules out of lattice rules of lower order.

Definition 3.23. *Suppose Q_1 and Q_2 are s-dimensional lattice rules having $\mathbf{0}$ as the only common quadrature point. If*

$$Q_1 f = \frac{1}{N_1} \sum_{j=0}^{N_1-1} f(\mathbf{x}_j) \tag{3.13}$$

and

$$Q_2 f = \frac{1}{N_2} \sum_{k=0}^{N_2-1} f(\mathbf{y}_k), \tag{3.14}$$

then the direct sum of Q_1 and Q_2, written $Q_1 \oplus Q_2$, is the s-dimensional rule defined by

$$(Q_1 \oplus Q_2)f := \frac{1}{N_1 N_2} \sum_{k=0}^{N_2-1} \sum_{j=0}^{N_1-1} f(\{\mathbf{x}_j + \mathbf{y}_k\}).$$

Example 3.24. *If*

$$Q_1 f = \frac{1}{n} \sum_{j=0}^{n-1} f\left(\frac{j}{n}, \frac{j}{n}\right)$$

and

$$Q_2 f = \frac{1}{m} \sum_{k=0}^{m-1} f\left(0, \frac{k}{m}\right),$$

then

$$(Q_1 \oplus Q_2)f = \frac{1}{nm} \sum_{k=0}^{m-1} \sum_{j=0}^{n-1} f\left(\frac{j}{n}, \left\{\frac{j}{n} + \frac{k}{m}\right\}\right).$$

As in this example, the direct sum of two lattice rules is itself a lattice rule.

Theorem 3.25. *Let Q_1 and Q_2 be s-dimensional lattice rules, of orders N_1 and N_2, respectively, having $\mathbf{0}$ as the only common quadrature point. The direct sum $Q_1 \oplus Q_2$ is a lattice rule of order $N_1 N_2$.*

Proof With Q_1 and Q_2 expressed as in (3.13) and (3.14), let $\mathcal{A}(Q_1)$ and $\mathcal{A}(Q_2)$ be the sets

$$\mathcal{A}(Q_1) = \{\mathbf{x}_j : \ 0 \leqslant j \leqslant N_1 - 1\}, \qquad \mathcal{A}(Q_2) = \{\mathbf{y}_k : \ 0 \leqslant k \leqslant N_2 - 1\}.$$

Theorem 3.4 tells us that $\mathcal{A}(Q_1)$ and $\mathcal{A}(Q_2)$ are groups under the group operation (3.4). Since they have only the point $\mathbf{0}$ in common, their direct sum (see Definition 3.8),

$$\mathcal{A}(Q_1) \oplus \mathcal{A}(Q_2) = \{\{\mathbf{x}_j + \mathbf{y}_k\} : \ 0 \leqslant j \leqslant N_1 - 1, 0 \leqslant k \leqslant N_2 - 1\},$$

is of order $N_1 N_2$. This is the set of quadrature points of $Q_1 \oplus Q_2$. It is also, clearly, the restriction to the half-open cube U^s of the integration lattice

$$\{\mathbf{x}_j + \mathbf{y}_k + \mathbf{z} : 0 \leqslant j \leqslant N_1 - 1, 0 \leqslant k \leqslant N_2 - 1, \mathbf{z} \in \mathbb{Z}^s\},$$

from which the result follows. ∎

Sometimes it is hard to determine whether two lattice rules have quadrature points other than $\mathbf{0}$ in common, but there is one circumstance in which we can be sure that they do not have any other point in common:

Theorem 3.26. *Two s-dimensional lattice rules of relatively prime order have $\mathbf{0}$ as their only common quadrature point.*

Proof Let the lattice rules be Q_1 and Q_2, and be of orders N_1 and N_2, respectively, where N_1 and N_2 are relatively prime. If $\mathcal{A}(Q_1)$ is the set of quadrature points of Q_1, then the order of every member of $\mathcal{A}(Q_1)$ divides N_1 and so is relatively prime to N_2. Thus a member of $\mathcal{A}(Q_1)$ cannot belong to the set of quadrature points of Q_2 unless it is of order 1, that is, unless it is the point $\mathbf{0}$. ∎

The theorem tells us that if Q_1 and Q_2 are of relatively prime order then the direct sum $Q_1 \oplus Q_2$ can be defined. The following converse is sometimes useful.

Theorem 3.27. *Let Q be a lattice rule of order $N = N_1 N_2$, where N_1 and N_2 are relatively prime. Then Q can be expressed, uniquely, in the form*

$$Q = Q_1 \oplus Q_2, \tag{3.15}$$

where Q_1 and Q_2 are lattice rules of orders N_1 and N_2 respectively.

Proof Let $\mathcal{A}(Q)$ be the group of quadrature points of Q as in Theorem 3.4, and let $\mathcal{A}_1$ be the subset of $\mathcal{A}(Q)$ consisting of the points whose orders divide N_1. In terms of the decomposition of $\mathcal{A}(Q)$ into Sylow subgroups as in (3.5), we see, because N_1 and N_2 are relatively prime, that $\mathcal{A}_1$ is the direct sum of the Sylow p subgroups of $\mathcal{A}(Q)$ for the primes p that divide N_1. It follows that $\mathcal{A}_1$ is of order N_1.

Similarly, $\mathcal{A}_2$, the subset of $\mathcal{A}(Q)$ consisting of the points whose orders divide N_2, is of order N_2. The direct sum $\mathcal{A}_1 \oplus \mathcal{A}_2$ is of order $N_1 N_2$, and so is the whole of $\mathcal{A}(Q)$. Letting Q_1 be the lattice rule with quadrature points $\mathcal{A}_1$, and Q_2 the lattice rule with quadrature points $\mathcal{A}_2$, the representation (3.15) now follows.

The uniqueness of the representation (3.15) follows from the fact that if $\mathbf{x} \in \mathcal{A}(Q_1)$ and $\mathbf{y} \in \mathcal{A}(Q_2)$, then (because N_1 and N_2 are relatively prime) the order of $\{\mathbf{x} + \mathbf{y}\}$ cannot divide N_1 unless $\mathbf{y} = \mathbf{0}$. Thus there are no potential candidates for membership of $\mathcal{A}(Q_1)$ other than those that already belong. ∎

Example 3.28. *Consider the* 12-*point lattice rule* Q *given by* (3.8). *To demonstrate explicitly its decomposition into the direct sum*

$$Q = Q_1 \oplus Q_2$$

of rules Q_1 *and* Q_2 *of orders* 4 *and* 3, *respectively, we may make use of the construction in the proof of the last theorem: the quadrature points of* Q_1 *and* Q_2 *are exactly the quadrature points of* Q *whose orders divide* 4 *and* 3 *respectively. We can read off the corresponding quadrature points from the full list of* 12 *quadrature points in Example* 3.16, *so obtaining*

$$Q_1 f = \frac{1}{4}\left[f(0,0) + f\left(\frac{1}{2},0\right) + f\left(\frac{1}{2},\frac{1}{2}\right) + f\left(0,\frac{1}{2}\right)\right],$$

$$Q_2 f = \frac{1}{3}\left[f(0,0) + f\left(\frac{1}{3},\frac{2}{3}\right) + f\left(\frac{2}{3},\frac{1}{3}\right)\right].$$

The corresponding direct sum representation of Q *is*

$$Q f = \frac{1}{12}\sum_{j=0}^{2}\sum_{k_2=0}^{1}\sum_{k_1=0}^{1} f\left(\left\{\left(\frac{k_1}{2},\frac{k_2}{2}\right) + j\left(\frac{1}{3},\frac{2}{3}\right)\right\}\right).$$

If the orders of the component lattice rules in a direct sum are relatively prime, then the rank and invariants of the direct sum are related in a simple way to those of the components.

Theorem 3.29. *Let* $Q = Q_1 \oplus Q_2$, *where* Q_1 *and* Q_2 *are lattice rules of orders* N_1 *and* N_2 *respectively. If* N_1 *and* N_2 *are relatively prime, then the rank of* Q *is the maximum of the ranks of* Q_1 *and* Q_2. *With the invariants of* Q_1 *and* Q_2 *written as* $n_1^{(1)}, n_2^{(1)}, \ldots, n_{r_1}^{(1)}$ *and* $n_1^{(2)}, n_2^{(2)}, \ldots, n_{r_2}^{(2)}$ *and with* r_1 *assumed to be* $\geqslant r_2$, *the invariants of* Q *are* $n_1, \ldots, n_{r_1}$, *where*

$$n_k = \begin{cases} n_k^{(1)} n_k^{(2)}, & k = 1, \ldots, r_2, \\ n_k^{(1)}, & k = r_2 + 1, \ldots, r_1. \end{cases}$$

Proof Since $N_1 = \prod_{k=1}^{r_1} n_k^{(1)}$ and $N_2 = \prod_{k=1}^{r_2} n_k^{(2)}$ are relatively prime, it follows that $n_k^{(1)}$ and $n_k^{(2)}$ are relatively prime for $k = 1, \ldots, r_2$.

Now observe from Theorem 3.22 that $\mathcal{A}(Q_1)$ and $\mathcal{A}(Q_2)$ can be expressed in the form

$$\mathcal{A}(Q_1) = \mathcal{D}_1^{(1)} \oplus \cdots \oplus \mathcal{D}_{r_1}^{(1)}, \qquad \mathcal{A}(Q_2) = \mathcal{D}_1^{(2)} \oplus \cdots \oplus \mathcal{D}_{r_2}^{(2)},$$

where $\mathcal{D}_k^{(1)}$ is a cyclic group of order $n_k^{(1)}$ and $\mathcal{D}_k^{(2)}$ is a cyclic group of order $n_k^{(2)}$. It follows that

$$\mathcal{A}(Q) = \mathcal{A}(Q_1) \oplus \mathcal{A}(Q_2) = \mathcal{D}_1 \oplus \cdots \oplus \mathcal{D}_{r_1}, \tag{3.16}$$

where

$$\mathcal{D}_k = \begin{cases} \mathcal{D}_k^{(1)} \oplus \mathcal{D}_k^{(2)}, & k = 1, \ldots, r_2, \\ \mathcal{D}_k^{(1)}, & k = r_2 + 1, \ldots, r_1. \end{cases}$$

Since $\mathcal{D}_k^{(1)}$ and $\mathcal{D}_k^{(2)}$ are of relatively prime orders, it follows from Lemma 3.18 that $\mathcal{D}_k$ is a cyclic group of order n_k. It is clear from the definition of the set $\{n_1, \ldots, n_{r_1}\}$ that n_{k+1} divides n_k for $k = 1, \ldots, r_1 - 1$ and that $n_{r_1} > 1$, so that it follows from (3.16) that $n_1, \ldots, n_{r_1}$ are the invariants of the rule Q, and that r_1 is its rank. ∎

Example 3.30. *If n and m in Example 3.24 are chosen relatively prime, then $Q := Q_1 \oplus Q_2$ can be written in the form*

$$Qf = \frac{1}{nm} \sum_{j=0}^{nm-1} f\left(\left\{ j\left(\frac{1}{n}, \frac{1}{n} + \frac{1}{m}\right) \right\}\right),$$

and so is a rank-1 rule. At the other extreme, if $n = m$ then the rule is just the product-rectangle rule with n^2 points, and so is manifestly a rule of rank 2 and invariants n, n. The latter example shows that the condition in Theorem 3.29 that N_1 and N_2 be relatively prime cannot be omitted.

Example 3.31. *The 12-point lattice rule (3.8) is manifestly a rule of rank 2 and invariants 6, 2. In Example 3.28 the rule is decomposed into two components, the first of which has rank 2 and invariants 2, 2, and the second has rank 1 and invariant 3. The relation between the invariants of the direct sum and its components conforms exactly with Theorem 3.29.*

3.4 Projections of lattice rules

Given a quadrature rule for the s-dimensional cube C^s, one way of obtaining a quadrature rule for the d-dimensional cube C^d with $d < s$ is simply to omit the last $s - d$ components of each quadrature point. We shall refer to a quadrature rule obtained in this way as a 'principal projection' of the original rule.

Definition 3.32. *Let d be any integer satisfying $1 \leqslant d \leqslant s$. The d-dimensional principal projection of a quadrature rule defined over the s-dimensional cube C^s is the d-dimensional rule obtained by omitting the last $s - d$ components of each quadrature point.*

More generally, we may omit any selection of $s - d$ components, not just those that happen to be last.

Definition 3.33. *Suppose d is an integer satisfying $1 \leqslant d \leqslant s$. A d-dimensional projection of a quadrature rule defined over the s-dimensional cube is a d-dimensional rule obtained by omitting a specified set of $s - d$ components of each quadrature point.*

Of course, two s-dimensional points which are distinct might no longer be distinct when some of the components are omitted. Thus a projected quadrature rule may have fewer distinct quadrature points than the quadrature rule from which it derives.

If the s-dimensional rule we start with is a lattice rule, then an important observation is that a d-dimensional projection is itself a lattice rule. This follows immediately from the obvious fact that the result of omitting some of the components of an s-dimensional integration lattice is still an integration lattice: it still satisfies the axioms of a lattice, and still contains integer vectors as a subset.

Another important property of the projections of lattice rules is that they inherit some of the structure of the parent lattice. This is made explicit in the following result from Sloan and Lyness (1989).

Theorem 3.34. *Let Q be an s-dimensional lattice rule having rank r and invariants $n_1, n_2, \ldots, n_r$. A d-dimensional projection of Q has rank $r' \leqslant r$ and invariants $n_1', n_2', \ldots, n_{r'}'$, where n_k' is a divisor of n_k for $1 \leqslant k \leqslant r'$.*

An immediate example may be instructive.

Example 3.35. *Let Q be the three-dimensional rule*

$$Qf = \frac{1}{12} \sum_{j_2=0}^{1} \sum_{j_1=0}^{5} f\left(\left\{\frac{j_1}{6}(3,1,1) + \frac{j_2}{2}(1,1,0)\right\}\right). \qquad (3.17)$$

Since this is a nonrepetitive rule in the canonical form of Theorem 3.2, we can immediately see that this rule has rank 2, and invariants 6 and 2. The two-dimensional principal projection is

$$\frac{1}{12} \sum_{j_2=0}^{1} \sum_{j_1=0}^{5} f\left(\left\{\frac{j_1}{6}(3,1) + \frac{j_2}{2}(1,1)\right\}\right).$$

This rule is repetitive: we easily find that there are only six distinct points, each of which occurs twice. It can also be written as

$$\frac{1}{6} \sum_{j=0}^{5} f\left(\left\{\frac{j}{6}(3,1)\right\}\right),$$

from which it is clear that the projected rule has rank 1 and sole invariant 6. Since $1 \leqslant 2$ and 6 divides 6, the properties asserted in Theorem 3.34 hold. As an exercise, the reader can check that the one-dimensional principal projection has rank 1 and sole invariant 2.

Rather than give a formal proof of Theorem 3.34, we prefer to try to explain the ideas behind it. (For an austere five-line proof, see Sloan and Lyness (1989).) The first step in the proof is to define $\mathcal{H}$, the subset of

$\mathcal{A}(Q)$ that projects on to the origin in the d-dimensional space. It can easily be shown that $\mathcal{H}$ is a group under the operation (3.4). Recall that in forming the d-dimensional projection of $\mathcal{A}(Q)$ we ignore $s - d$ components of each point of $\mathcal{A}(Q)$—or, what is the same thing, we treat as equivalent points of $\mathcal{A}(Q)$ which differ by an element of $\mathcal{H}$, and which therefore become indistinguishable when the d-dimensional projection is taken. Technically, forming the d-dimensional projection of $\mathcal{A}(Q)$ is equivalent to forming the 'factor group' $\mathcal{A}(Q)/\mathcal{H}$, the elements of which are subsets of $\mathcal{A}(Q)$ (known in group theory as 'cosets') in which each member differs from every other member of that coset merely by an element of $\mathcal{H}$. The point of all this is that *every member of a given coset projects on to the same d-dimensional point*. And *different cosets project on to different d-dimensional points*. Each coset of $\mathcal{H}$ contains the same number of elements.

Example 3.36. *Let $\mathcal{A}(Q)$ be the set of quadrature points for the lattice rule in (3.17) so that*

$$
\mathcal{A}(Q) = \left\{ (0,0,0), \left(\frac{1}{2},\frac{1}{6},\frac{1}{6}\right), \left(0,\frac{1}{3},\frac{1}{3}\right), \left(\frac{1}{2},\frac{1}{2},\frac{1}{2}\right), \right.
$$
$$
\left(0,\frac{2}{3},\frac{2}{3}\right), \left(\frac{1}{2},\frac{5}{6},\frac{5}{6}\right), \left(\frac{1}{2},\frac{1}{2},0\right), \left(0,\frac{2}{3},\frac{1}{6}\right),
$$
$$
\left. \left(\frac{1}{2},\frac{5}{6},\frac{1}{3}\right), \left(0,0,\frac{1}{2}\right), \left(\frac{1}{2},\frac{1}{6},\frac{2}{3}\right), \left(0,\frac{1}{3},\frac{5}{6}\right) \right\}.
$$

The subset of $\mathcal{A}(Q)$ that projects on to $(0,0)$ when the third component is ignored is

$$
\mathcal{H} = \left\{ (0,0,0), \left(0,0,\frac{1}{2}\right) \right\}.
$$

The cosets of $\mathcal{H}$ in $\mathcal{A}(Q)$ (or, equivalently, the elements of $\mathcal{A}(Q)/\mathcal{H}$) are

$$
\left\{ (0,0,0), \left(0,0,\frac{1}{2}\right) \right\}, \left\{ \left(\frac{1}{2},\frac{1}{6},\frac{1}{6}\right), \left(\frac{1}{2},\frac{1}{6},\frac{2}{3}\right) \right\},
$$
$$
\left\{ \left(0,\frac{1}{3},\frac{1}{3}\right), \left(0,\frac{1}{3},\frac{5}{6}\right) \right\}, \left\{ \left(\frac{1}{2},\frac{1}{2},\frac{1}{2}\right), \left(\frac{1}{2},\frac{1}{2},0\right) \right\},
$$
$$
\left\{ \left(0,\frac{2}{3},\frac{2}{3}\right), \left(0,\frac{2}{3},\frac{1}{6}\right) \right\}, \left\{ \left(\frac{1}{2},\frac{5}{6},\frac{5}{6}\right), \left(\frac{1}{2},\frac{5}{6},\frac{1}{3}\right) \right\},
$$

each coset consisting of points that become identical when the third component is ignored. The six cosets project on to six different two-dimensional points, namely

$$
(0,0), \left(\frac{1}{2},\frac{1}{6}\right), \left(0,\frac{1}{3}\right), \left(\frac{1}{2},\frac{1}{2}\right), \left(0,\frac{2}{3}\right), \left(\frac{1}{2},\frac{5}{6}\right),
$$

in agreement with the final expression in Example 3.35.

It follows that the properties of principal projections are just the properties of factor groups $\mathcal{A}(Q)/\mathcal{H}$. Now we use the known result that *every factor group of a finite abelian group $\mathcal{G}$ is isomorphic to some subgroup of $\mathcal{G}$.* (In the case of Example 3.36 the subgroup of $\mathcal{A}(Q)$ consisting of the first 6 of the listed elements of $\mathcal{A}(Q)$ is clearly isomorphic to $\mathcal{A}(Q)/\mathcal{H}$.) This allows us to make use of a known structure theorem for the subgroups of a finite abelian group:

Theorem 3.37. *A subgroup $\mathcal{G}'$ of a finite abelian group having rank r and invariants $n_1, n_2, \ldots, n_r$ has rank $r' \leqslant r$ and invariants $n'_1, n'_2, \ldots, n'_{r'}$, where n'_k divides n_k for $1 \leqslant k \leqslant r'$.*

The proof of this theorem is by way of the same decomposition into prime-power cyclic groups discussed in the preceding section. Theorem 3.34 may now be seen as an immediate application of this last theorem. Further discussion can be found in Sloan and Lyness (1989).

Properties of projections of lattice rules can play an important role even when we are only interested in s-dimensional rules: suppose that one term in a complicated integrand is a function only of $x_1, \ldots, x_d$, with $d < s$. Then as far as that term is concerned, the lattice rule has exactly the same effect as its d-dimensional principal projection.

For this reason we might be interested to have d-dimensional projections that are as rich as possible. The main importance of Theorem 3.34 may be that it allows us to understand what the possibilities really are. We shall exploit this understanding in the next section.

3.5 Projection-regular rules

In this section we consider lattice rules whose principal projections are as rich as possible, given the constraints imposed by Theorem 3.34. First, however, it is convenient to extend the notion of the 'invariants' of a lattice rule of rank r (where r, we recall, is the minimum number of sums required to write the lattice rule in the form (3.10)) by also defining $n_{r+1} = \cdots = n_s = 1$. The canonical form (3.3) can then be extended artificially to the s-fold sum

$$Qf = \frac{1}{n_1 n_2 \ldots n_s} \sum_{j_s=0}^{n_s-1} \cdots \sum_{j_1=0}^{n_1-1} f\left(\left\{\frac{j_1}{n_1}\mathbf{z}_1 + \cdots + \frac{j_s}{n_s}\mathbf{z}_s\right\}\right), \qquad (3.18)$$

where $\mathbf{z}_{r+1}, \ldots, \mathbf{z}_s$ are any integer vectors. We shall say that a rule in this form is in 'extended canonical form'. Note that the additional terms $j_{r+1}\mathbf{z}_{r+1}/n_{r+1}, \ldots, j_s\mathbf{z}_s/n_s$, being just vectors of integers, do not change the argument of f at all! Note, too, that the order is still $N = n_1 n_2 \cdots n_s$. The usefulness of this device is just that it makes some results a little easier to state. For example, as a corollary to Theorem 3.34 we have:

Corollary The order of a d-dimensional projection of an s-dimensional lattice rule with invariants $n_1, \ldots, n_s$ is a divisor of $n_1 n_2 \cdots n_d$.

Theorem 3.34 leads us to define a class of 'projection-regular' lattice rules, in which all the principal projections have maximum possible order.

Definition 3.38. *Suppose Q is an s-dimensional lattice rule with invariants $n_1, \ldots, n_s$. The rule is* projection regular *if for $1 \leqslant d \leqslant s$ the principal projection Q_d has order $n_1 n_2 \cdots n_d$, and hence invariants $n_1, \ldots, n_d$.*

Example 3.39. *The three-dimensional lattice rule*

$$Qf = \frac{1}{12} \sum_{j_2=0}^{1} \sum_{j_1=0}^{5} f\left(\left\{\frac{j_1}{6}(1,2,1) + \frac{j_2}{2}(1,1,1)\right\}\right) \qquad (3.19)$$

is a projection-regular rule, with rank 2 and invariants 6, 2, 1 (or just 6, 2). Its two-dimensional principal projection (which coincides with (3.8)) has invariants 6, 2, and its one-dimensional principal projection has the sole invariant 6.

Example 3.40. *The three-dimensional rule (3.17) with invariants 6, 2, 1 is not projection regular—neither its two-dimensional nor its one-dimensional principal projections are of the 'correct' order (namely, 12 and 6 respectively).*

Theorem 3.41 below establishes that the direct sum of two projection-regular lattice rules of relatively prime orders is itself projection regular. The first part of the theorem is just a restatement of Theorem 3.29 in the language of the extended invariants $n_1, \ldots, n_s$, in which some of them may have the value 1.

Theorem 3.41. *Let $Q = Q_1 \oplus Q_2$ be the direct sum of lattice rules Q_1 and Q_2 of relatively prime orders.*

(i) *If Q_1 and Q_2 have extended invariants $n_1^{(1)}, \ldots, n_s^{(1)}$ and $n_1^{(2)}, \ldots, n_s^{(2)}$, respectively, then the invariants of Q are $n_1^{(1)} n_1^{(2)}, \ldots, n_s^{(1)} n_s^{(2)}$.*

(ii) *Q is projection regular if and only if Q_1 and Q_2 are projection regular.*

Proof The proof of part (i) follows from Theorem 3.29. Part (ii) follows from part (i) and Theorem 3.34: the leading d invariants of Q can take their maximum values $n_1^{(1)} n_1^{(2)}, \ldots, n_d^{(1)} n_d^{(2)}$ if and only if the leading d invariants of Q_1 and Q_2 also take their maximum values. ∎

One reason for considering projection-regular rules is that for this class of rules a rather precise representation is available. We recall that in the canonical form (3.3) or the extended canonical form (3.18) the vectors $\mathbf{z}_j$ are not uniquely determined. However, for the class of projection-regular rules Sloan and Lyness (1990) have shown that additional constraints can be imposed so that these vectors become unique. The remainder of this section will be devoted to this topic.

Before proceeding further, it is convenient to define the 'Z-matrix' of a lattice rule expressed in the extended canonical form (3.18).

Definition 3.42. *The Z-matrix of the extended canonical form (3.18) is the $s \times s$ matrix given by*

$$Z = \begin{bmatrix} \mathbf{z}_1 \\ \vdots \\ \mathbf{z}_s \end{bmatrix}.$$

Thus z_{kp}, the element of Z in the (k,p)th position, is the pth component of $\mathbf{z}_k$.

Example 3.43. *The Z-matrix of the rule (3.19) is*

$$Z = \begin{bmatrix} 1 & 2 & 1 \\ 1 & 1 & 1 \\ a & b & c \end{bmatrix},$$

where a, b, and c are arbitrary integers.

Before concerning ourselves with uniqueness questions, we first have the following result from Sloan and Lyness (1990).

Theorem 3.44. *If the lattice rule Q is projection regular, then it may be expressed in extended canonical form with an upper unit triangular Z-matrix.*

An upper unit triangular matrix is an upper triangular matrix with ones along the main diagonal. We shall omit the details of the proof, but the idea behind it is easy to understand. One starts with the original Z-matrix of the given projection-regular lattice rule, and then uses row transformations in a manner reminiscent of Gaussian elimination to change Z progressively until finally it becomes upper unit triangular, with the transformations being chosen so that the lattice rule is unchanged. The projection-regularity assumption prevents the occurrence of unwelcome zeros. For later use, we record the set of transformations used by Sloan and Lyness (1990), in the following easily verified proposition.

Proposition 3.45. *The lattice rule Q given by (3.18) is unchanged if, for given k satisfying $1 \leqslant k \leqslant s$, the vector $\mathbf{z}_k$ is replaced by any of:*

(i) $\mathbf{z}'_k = \ell \mathbf{z}_k$ *if $\ell \in \mathbb{Z}$, and ℓ and n_k are relatively prime.*

(ii) $\mathbf{z}'_k = \mathbf{z}_k + n_k \mathbf{w}$, *where $\mathbf{w} \in \mathbb{Z}^s$.*

(iii) $\mathbf{z}'_k = \mathbf{z}_k + \ell \dfrac{n_k}{n_p} \mathbf{z}_p$, *if $p \neq k$, $\ell \in \mathbb{Z}$ and n_p divides ℓn_k.*

The converse of Theorem 3.44 is also true.

Theorem 3.46. *If the lattice rule Q can be expressed in the form* (3.18) *with $n_1, \ldots, n_s$ satisfying*

$$n_{k+1} \text{ divides } n_k, \quad 1 \leqslant k \leqslant s - 1, \tag{3.20}$$

and with the Z-matrix upper unit triangular, then Q is a projection-regular rule of order $n_1 n_2 \cdots n_s$.

Proof By omitting the last $s - d$ components of the quadrature points of Q, we see that all the quadrature points of the principal projection Q_d belong to the set

$$\left\{ \left\{ \sum_{k=1}^{s} \frac{j_k}{n_k} (z_{k1}, \ldots, z_{kd}) \right\} : 0 \leqslant j_k \leqslant n_k - 1, \ 1 \leqslant k \leqslant s \right\}. \tag{3.21}$$

Since the Z-matrix is upper triangular, the first d components of each of $\mathbf{z}_{d+1}, \ldots, \mathbf{z}_s$ must be zero, that is, $z_{kp} = 0$ for $d + 1 \leqslant k \leqslant s$ and $1 \leqslant p \leqslant d$. Thus the set given in (3.21) may be written as

$$\left\{ \left\{ \sum_{k=1}^{d} \frac{j_k}{n_k} (z_{k1}, \ldots, z_{kd}) \right\} : 0 \leqslant j_k \leqslant n_k - 1, \ 1 \leqslant k \leqslant d \right\}. \tag{3.22}$$

Since this set taken at face value contains $n_1 n_2 \cdots n_d$ elements, we can prove that the lattice rule is projection regular by proving that each of these elements is distinct, so that the order of Q_d is exactly $n_1 n_2 \cdots n_d$ for $d = 1, \ldots, s$.

To do this, we observe that, because Z is upper unit triangular, if for a fixed value of k satisfying $1 \leqslant k \leqslant d$ we set $j_p = 0$ for $p \neq k$ in (3.22), then we obtain the subset of quadrature points given by

$$\left\{ \left\{ \frac{j_k}{n_k} (0, \ldots, 0, 1, z_{k,k+1}, \ldots, z_{kd}) \right\} : 0 \leqslant j_k \leqslant n_k - 1 \right\},$$

where the 1 occurs in the kth position. The kth component of the j_kth vector in this set is j_k/n_k, and so is different for each value of j_k, and hence the n_k quadrature points in this subset are distinct. Moreover, if we start with $k = d$ and then reduce k progressively, the n_k quadrature points for a particular value of k have clearly not been encountered previously. Letting k range from d to 1, we conclude that the set in (3.22) contains $n_1 n_2 \cdots n_d$ distinct quadrature points, from which we conclude that Q is projection regular. $\blacksquare$

Taken together, Theorems 3.44 and 3.46 show that a lattice rule Q is projection regular if and only if there exists an extended canonical form for which the Z matrix is upper unit triangular. This leads to a 'standard form' for projection-regular rules, in which all the elements of Z are uniquely determined.

Definition 3.47. *The lattice rule Q given by (3.18) is in standard form if (3.20) holds and the elements of its Z-matrix satisfy*

$$z_{kp} = 0, \qquad 1 \leqslant p < k \leqslant s,$$

$$z_{kk} = 1, \qquad 1 \leqslant k \leqslant s,$$

$$0 \leqslant z_{kp} < \frac{n_k}{n_p}, \qquad 1 \leqslant k < p \leqslant s. \tag{3.23}$$

It follows from Theorem 3.46 that any lattice rule Q in standard form must be projection regular.

Example 3.48. *The rule*

$$Qf = \frac{1}{12} \sum_{j_2=0}^{1} \sum_{j_1=0}^{5} f\left(\left\{ \frac{j_1}{6}(1,2,1) + \frac{j_2}{2}(0,1,0) + \frac{0}{1}(0,0,1) \right\}\right)$$

is in standard form, having invariants 6, 2, 1 and Z-matrix

$$Z = \begin{bmatrix} 1 & 2 & 1 \\ 0 & 1 & 0 \\ 0 & 0 & 1 \end{bmatrix}.$$

The rule is a projection-regular rule, which coincides with the rule (3.19).

We now come to the main result of this section.

Theorem 3.49. *A projection-regular lattice rule can be expressed in standard form. The representation in standard form is unique.*

Proof To prove Theorem 3.49, we need to show that the Z-matrix for a projection-regular lattice rule can always be chosen to be in standard form and that this Z-matrix is unique.

It follows from Theorem 3.44 that the Z-matrix can be assumed to be upper unit triangular. If it is not already in standard form, then we shall show that it can be transformed into standard form by using a sequence of transformations

$$\mathbf{z}'_k = \mathbf{z}_k - \left\lfloor \frac{z_{kp}n_p}{n_k} \right\rfloor \frac{n_k}{n_p} \mathbf{z}_p, \quad k < p, \tag{3.24}$$

where $\lfloor a \rfloor$ denotes the integer part of a. Since n_p divides n_k, Proposition 3.45(iii) shows that such a transformation leaves the lattice rule Q unchanged. The transformation (3.24) affects only $\mathbf{z}_k$, and hence only the kth row of Z. Moreover, since the first $p-1$ components of $\mathbf{z}_p$ are zero, the transformation leaves the first $p-1$ components of $\mathbf{z}_k$ unchanged, but generally alters the remaining components. In particular, since $z_{pp} = 1$, we see that z_{kp} is replaced by

$$z'_{kp} = z_{kp} - \left\lfloor \frac{z_{kp} n_p}{n_k} \right\rfloor \frac{n_k}{n_p},$$

which satisfies (3.23). Once z_{kp} has been replaced by z'_{kp}, any further transformations of the form (3.24) must be ordered in such a way that this new element z'_{kp} is not altered again. This property clearly holds if we deal successively with $\mathbf{z}_1, \ldots, \mathbf{z}_{s-1}$, and in each vector $\mathbf{z}_k$ alter the off-diagonal components z_{kp}, $p > k$, in order of increasing p.

We shall now use induction to prove that a Z-matrix in standard form is unique. Suppose Z and Z' are two alternative Z-matrices for the rule Q, both in standard form. Then the first columns of Z and Z' are the same (the first component is 1 and all others are zero).

Now assume that the first $d-1$ columns of Z and Z' are the same. Given any integer p satisfying $1 \leqslant p \leqslant d-1$, (3.18) tells us that the vectors $\mathbf{z}_p/n_p$ and $\mathbf{z}'_p/n_p$ belong to the integration lattice corresponding to Q. From the properties of a lattice, the difference $(\mathbf{z}_p - \mathbf{z}'_p)/n_p$ does also. If we retain just the first d components of this difference vector, and take the absolute value, we obtain the d-dimensional vector

$$\left(0, \ldots, 0, \frac{|z_{pd} - z'_{pd}|}{n_p} \right), \tag{3.25}$$

which by construction belongs to the lattice corresponding to the d-dimensional principal projection Q_d. Thus there exist integers $j_1, \ldots, j_d$ with $0 \leqslant j_k \leqslant n_k - 1$ such that

$$\left\{ \left(0, \ldots, 0, \frac{|z_{pd} - z'_{pd}|}{n_p} \right) \right\} = \left\{ \frac{j_1}{n_1}(1, z_{12}, \ldots, z_{1d}) + \cdots + \frac{j_d}{n_d}(0, \ldots, 0, 1) \right\},$$

giving successively $j_1 = 0, \ldots, j_{d-1} = 0$, and finally

$$\left\{ \frac{|z_{pd} - z'_{pd}|}{n_p} \right\} = \left\{ \frac{j_d}{n_d} \right\} = \frac{j_d}{n_d}. \tag{3.26}$$

Now by (3.23) both z_{pd} and z'_{pd} are in the interval $[0, n_p/n_d)$, thus it follows that

$$\left\{ \frac{|z_{pd} - z'_{pd}|}{n_p} \right\} = \frac{|z_{pd} - z'_{pd}|}{n_p} < \frac{1}{n_d}, \tag{3.27}$$

so that (3.26) can be satisfied only if $j_d = 0$, and hence $z_{pd} = z'_{pd}$. Thus if the first $d-1$ columns of Z and Z' coincide, then so does their dth column. It follows by induction that Z and Z' must be the same matrix. Thus we conclude that the Z-matrix in standard form is unique. ∎

Since the standard form is unique, it is possible to work out the number of different projection-regular lattice rules having a given set of (extended) invariants $n_1, \ldots, n_s$.

Corollary Given the invariants $n_1, \ldots, n_s$, the number of different projection-regular lattice rules having such invariants is

$$n_1^{s-1} n_2^{s-3} \cdots n_{s-1}^{3-s} n_s^{1-s}.$$

Proof For the Z-matrix in standard form, there are n_k/n_p possible choices of z_{kp} for $1 \leqslant k < p \leqslant s$. Thus by looking at the upper triangular elements of each of the columns of the Z-matrix in turn, we see that the number of different lattice rules is given by

$$
\begin{aligned}
&\frac{n_1}{n_2} \times \left(\frac{n_1}{n_3} \times \frac{n_2}{n_3} \right) \times \cdots \times \left(\frac{n_1}{n_s} \times \cdots \times \frac{n_{s-1}}{n_s} \right) \\
=\; & \frac{n_1}{n_2} \times \frac{n_1 n_2}{n_3^2} \times \cdots \times \frac{n_1 n_2 \cdots n_{s-1}}{n_s^{s-1}} \\
=\; & \frac{n_1^{s-1} n_2^{s-2} \cdots n_{s-2}^2 n_{s-1}}{n_2 n_3^2 \cdots n_{s-1}^{s-2} n_s^{s-1}} \\
=\; & n_1^{s-1} n_2^{s-3} \cdots n_{s-1}^{3-s} n_s^{1-s},
\end{aligned}
$$

which completes the proof. ∎

3.6 Results for rank-2 rules

Later we shall see (in Section 5.4) that there may sometimes be advantages in expressing certain rules of rank 2 not in the standard two-sum form, but in a three-sum form, as in (3.29) below. The results given below were obtained by Lyness and Sloan (1989), but here they are obtained differently as an exercise on 'direct sums' of lattice rules, as defined in Definition 3.23.

By definition, a lattice rule of rank 2 has two invariants (or, from the point of view of extended invariants, two invariants that exceed 1), which we may write as nm and n. Expressed in the canonical form, a rank-2 lattice rule with such invariants can be written as

$$Qf = \frac{1}{n^2 m} \sum_{j_2=0}^{n-1} \sum_{j_1=0}^{nm-1} f\left(\left\{ \frac{j_1}{nm} \mathbf{z}_1 + \frac{j_2}{n} \mathbf{z}_2 \right\} \right), \tag{3.28}$$

in which each of the $n^2 m$ points is distinct.

Theorem 3.50. *If n and m are relatively prime, then a rank-2 lattice rule with invariants nm and n may be expressed in the form*

$$Qf = \frac{1}{n^2 m} \sum_{k_2=0}^{n-1} \sum_{k_1=0}^{n-1} \sum_{j=0}^{m-1} f\left(\left\{\frac{j}{m}\mathbf{z} + \frac{k_1}{n}\mathbf{y}_1 + \frac{k_2}{n}\mathbf{y}_2\right\}\right), \qquad (3.29)$$

with $\mathbf{z}, \mathbf{y}_1, \mathbf{y}_2 \in \mathbb{Z}^s$.

Proof Since n^2 and m are relatively prime, it follows from Theorem 3.27 that Q can be expressed as $Q = Q_1 \oplus Q_2$, where Q_1 is of order n^2 and Q_2 is of order m. Theorem 3.29 shows that Q_1 must be of rank 2 and have invariants n, n, while Q_2 must be of rank 1 and have invariant m. Thus Q_1 is expressible in the form

$$Q_1 f = \frac{1}{n^2} \sum_{k_2=0}^{n-1} \sum_{k_1=0}^{n-1} f\left(\left\{\frac{k_1}{n}\mathbf{y}_1 + \frac{k_2}{n}\mathbf{y}_2\right\}\right), \qquad (3.30)$$

and Q_2 in the form

$$Q_2 f = \frac{1}{m} \sum_{j=0}^{m-1} f\left(\left\{\frac{j}{m}\mathbf{z}\right\}\right), \qquad (3.31)$$

with $\mathbf{y}_1, \mathbf{y}_2, \mathbf{z} \in \mathbb{Z}^s$. The form (3.29) is the representation of Q as the direct sum of Q_1 and Q_2. ∎

Under certain conditions on the leading components of $\mathbf{z}$, $\mathbf{y}_1$, and $\mathbf{y}_2$, the lattice rule given in (3.29) is projection regular.

Theorem 3.51. *Suppose n and m are relatively prime. Let Q be an s-dimensional lattice rule given in the form (3.29), with*

$$z_1 = 1, \qquad y_{11} = 1, \qquad y_{21} = 0, \qquad y_{22} = 1, \qquad (3.32)$$

where the pth component of $\mathbf{y}_k$ is denoted by y_{kp}, $1 \leqslant k \leqslant 2$, $1 \leqslant p \leqslant s$. Then Q is a projection-regular rank-2 rule of order $n^2 m$ having invariants nm and n.

Proof Since n^2 and m are relatively prime, the rule (3.29) is the direct sum of the rules (3.30) and (3.31). The assumption (3.32) ensures that the leading entry of the first row of the Z-matrix for the n^2-point rule (3.30) is 1, and the first two entries of the second row are 0 and 1. Since the remaining rows of the Z-matrix may be chosen arbitrarily, we may choose the Z-matrix to be upper unit triangular, and deduce from Theorem 3.46 that the rule (3.30) is projection regular. Similarly, but more simply, the rule (3.31) is projection regular. Since n^2 and m are relatively prime, it now follows from Theorem 3.41 that the rule (3.29) is projection regular, and has invariants nm and n. ∎

The final theorem gives the converse of Theorem 3.51. The significance of this result from Lyness and Sloan (1989) lies in the fact that it shows that the vectors $\mathbf{z}$, $\mathbf{y}_1$, and $\mathbf{y}_2$ in the three-sum form are uniquely determined, if suitably constrained.

Theorem 3.52. *Let Q be an s-dimensional lattice rule with rank 2 and invariants nm and n, where n and m are relatively prime. If Q is projection regular, then it can be expressed in the form (3.29), with the leading components of $\mathbf{z}$, $\mathbf{y}_1$, and $\mathbf{y}_2$ satisfying*

$$z_1 = 1, \qquad y_{11} = 1, \qquad y_{12} = 0, \qquad y_{21} = 0, \qquad y_{22} = 1. \qquad (3.33)$$

The remaining components of $\mathbf{z}$, $\mathbf{y}_1$, and $\mathbf{y}_2$ are uniquely determined modulo m, n, and n respectively. Hence they can be chosen uniquely to satisfy

$$0 \leqslant z_p < m, \; 2 \leqslant p \leqslant s, \qquad 0 \leqslant y_{kp} < n \text{ for } k = 1, 2, \; 3 \leqslant p \leqslant s. \qquad (3.34)$$

Proof Since n and m are relatively prime, Theorem 3.50 tells us that the rank-2 rule Q with invariants nm and n is the direct sum of rules Q_1 and Q_2 of the forms (3.30) and (3.31). Since Q is projection regular, Theorem 3.41(ii) says that Q_1 and Q_2 are projection regular. Theorem 3.49 asserts that Q_1 and Q_2 can be expressed, uniquely, in standard form.

Since the invariants of Q_1 are n, n and the invariant of Q_2 is m, the Z-matrices of Q_1 and Q_2 are in standard form only if (3.33) and (3.34) are satisfied, and if the remaining rows of each (which are in any case of no consequence) are appropriately chosen. ∎

4

Rank-1 rules—the method of good lattice points

4.1 Introduction

In this chapter we look at rank-1 rules, which are rules of the form

$$Q(\mathbf{z}, N)f = \frac{1}{N} \sum_{j=0}^{N-1} f\left(\left\{\frac{j}{N}\mathbf{z}\right\}\right), \qquad (4.1)$$

where $\mathbf{z}$ is an s-dimensional integer vector having no factor in common with N. Such rules were first introduced by Korobov (1959), Bahvalov (1959), and Hlawka (1962). They are often referred to as number-theoretic rules since their accuracy is related to the number-theoretic properties of N and the components of $\mathbf{z}$ (for example, see Korobov 1963 or Hua and Wang 1981). The method has a large literature (surveys may be found in Haber 1970 and Niederreiter 1978a, 1988, 1992b), in which it is also known as the method of good lattice points. Rank-1 rules with well-chosen $\mathbf{z}$ are attractive to use in practical calculations, because the coding of (4.1), once $\mathbf{z}$ is available, is independent of the dimension s. An implementation of the method may be found in the NAG numerical software library (Numerical Algorithms Group 1991). The NAG software contains an error estimator that makes use of a randomization technique to be explained later in this chapter.

In the next section we look at possible criteria for measuring the 'goodness' of a rank-1 rule. One such criterion is a quantity $P_\alpha(\mathbf{z}, N)$, which bounds the quadrature error for all test functions belonging to a certain smoothness class. Most of the 'good' choices of $\mathbf{z}$ that are used in practice have been obtained by minimizing a quantity such as $P_\alpha(\mathbf{z}, N)$ over some family of possible choices of $\mathbf{z}$.

Only in the special case of two dimensions are explicit lattice rule constructions known that achieve the best rate of convergence. These are the celebrated Fibonacci rank-1 lattice rules, which we shall describe in Section 4.3.

In higher dimensions existence theorems play an important theoretical role, because of the absence of explicit constructions. The main task in this chapter is to establish theoretically the existence of good choices of $\mathbf{z}$. More precisely, the aim is to establish the existence of a sequence $\mathbf{z}(N)$ ('good lattice points') for which $P_\alpha(\mathbf{z}, N)$ converges to zero in an appropriately fast way as $N \to \infty$. This task is stated mathematically, in terms of the order of convergence of $P_\alpha(\mathbf{z}, N)$ as $N \to \infty$, in Section 4.4. Then in Section 4.5 we establish the existence of a sequence $\mathbf{z}(N)$ which achieves this order, by means of an appropriate averaging technique.

The order of convergence of $P_\alpha(\mathbf{z}, N)$ is not the only matter to be considered: also important, as with any error bound, is the size of the associated constant factors. The constant factors given here, taken from Disney and Sloan (1991), represent a significant reduction from those known previously.

The key to error bounds of rank-1 rules, as for all lattice rules, is the representation (2.20) of the error in terms of the Fourier coefficients of f. In the particular case of rank-1 rules it is easy to see that the dual lattice (2.19) becomes

$$L^\perp = \{\mathbf{h} \in \mathbb{Z}^s : \mathbf{h} \cdot \mathbf{z} \equiv 0 \, (\mathrm{mod} \, N)\}, \qquad (4.2)$$

so that for the rank-1 rules (2.20) becomes

$$Q(\mathbf{z}, N)f - If = \sideset{}{'}\sum_{\mathbf{h}\cdot\mathbf{z}\equiv 0 \,(\mathrm{mod} \, N)} \hat{f}(\mathbf{h}). \qquad (4.3)$$

4.2   Criteria for measuring 'goodness'

To use the rule (4.1) in practice, one needs to choose a vector $\mathbf{z}$ which is 'good' in some sense. In this section we shall look at several measures of the goodness of a rank-1 rule. All are rather standard in the number-theoretic literature associated with such rules. While stated here for rank-1 rules, all extend in a natural way to general lattice rules (see Section 5.2). A more complete discussion of possible criteria for assessing lattice rules may be found in Lyness (1988, 1989) and in Beckers (1992).

For $c > 0$ and fixed $\alpha > 1$, let $E_\alpha(c)$ be the class of functions f whose Fourier coefficients satisfy

$$|\hat{f}(\mathbf{h})| \leqslant \frac{c}{(\overline{h}_1 \overline{h}_2 \cdots \overline{h}_s)^\alpha},$$

where

$$\overline{h} = \max(1, |h|). \qquad (4.4)$$

The rate of decay of the Fourier coefficients of a function is related to the smoothness of the function, so $E_\alpha(c)$ is essentially a class of functions of a certain smoothness. The following well-known result, proved by Zaremba (1968), provides a useful sufficient condition for membership of this class when $\alpha > 1$ is an integer.

Theorem 4.1. *Suppose $\alpha > 1$ is an integer. If f is a one-periodic function on $\mathbb{R}^s$ whose partial derivatives*

$$\frac{\partial^{q_1 + \cdots + q_s} f}{\partial x_1^{q_1} \cdots \partial x_s^{q_s}}, \quad 0 \leqslant q_k \leqslant \alpha - 1, \quad 1 \leqslant k \leqslant s,$$

exist and are of bounded variation on C^s in the sense of Hardy and Krause, then there exists a $c > 0$ for which $f \in E_\alpha(c)$.

For the definition of the variation of a function in the sense of Hardy and Krause, we refer to Niederreiter (1978a). A sufficient condition for a function g defined on C^s to be of bounded variation in the sense of Hardy and Krause is that each of the mixed first partial derivatives

$$\frac{\partial^{q_1 + \cdots + q_s} g}{\partial x_1^{q_1} \cdots \partial x_s^{q_s}}, \quad 0 \leqslant q_k \leqslant 1, \quad 1 \leqslant k \leqslant s,$$

be continuous on C^s. This is not a necessary condition, as the variation is also bounded for some discontinuous functions, for example

$$g(x_1, x_2) = \left\{ \begin{array}{ll} 0, & 0 \leqslant x_1 < a, \\ 1, & a \leqslant x_1 \leqslant 1, \end{array} \right. \quad 0 \leqslant x_2 \leqslant 1,$$

where $0 < a < 1$. On the other hand, for other equally simple functions such as

$$g(x_1, x_2) = \left\{ \begin{array}{ll} 0, & 0 \leqslant x_1 + x_2 < 1, \\ 1, & 1 \leqslant x_1 + x_2 \leqslant 2, \end{array} \right. \quad 0 \leqslant x_1, x_2 \leqslant 1,$$

the variation is infinite. Thus as a simple corollary of Theorem 4.1 we have:

Corollary Suppose $\alpha > 1$ is an integer. If f is a one-periodic function on $\mathbb{R}^s$ whose partial derivatives

$$\frac{\partial^{q_1 + \cdots + q_s} f}{\partial x_1^{q_1} \cdots \partial x_s^{q_s}}, \quad 0 \leqslant q_k \leqslant \alpha, \quad 1 \leqslant k \leqslant s,$$

are continuous on C^s, then there exists a $c > 0$ for which $f \in E_\alpha(c)$.

For the rank-1 rule (4.1) applied to $f \in E_\alpha(c)$, there exists a simple error bound: it follows from (4.3) that for $f \in E_\alpha(c)$ the rank-1 rule $Q(\mathbf{z}, N)$ has an error satisfying

$$|Q(\mathbf{z}, N)f - If| \leqslant c \sum_{\substack{\mathbf{h} \cdot \mathbf{z} \equiv 0 \,(\mathrm{mod}\, N)}}' \frac{1}{(\overline{h}_1 \overline{h}_2 \cdots \overline{h}_s)^\alpha}. \tag{4.5}$$

Now let us define a function f_α, with $\alpha > 1$, by

$$f_\alpha(\mathbf{x}) := \sum_{\mathbf{h} \in \mathbb{Z}^s} \frac{1}{(\overline{h}_1 \overline{h}_2 \cdots \overline{h}_s)^\alpha} e^{2\pi i \mathbf{h} \cdot \mathbf{x}}. \tag{4.6}$$

Then $f_\alpha \in E_\alpha(1)$ and $If_\alpha = 1$. The function f_α is a worst possible function in the class $E_\alpha(1)$, because for this function the inequality (4.5) becomes an equality: from (4.3) we have

$$Q(\mathbf{z}, N)f_\alpha - 1 = \sideset{}{'}\sum_{\mathbf{h}\cdot\mathbf{z}\equiv 0\,(\mathrm{mod}\,N)} \frac{1}{(\bar{h}_1\bar{h}_2\cdots\bar{h}_s)^\alpha}. \tag{4.7}$$

Thus we define

$$P_\alpha(\mathbf{z}, N) = P_\alpha(Q(\mathbf{z}, N)) := \sideset{}{'}\sum_{\mathbf{h}\cdot\mathbf{z}\equiv 0\,(\mathrm{mod}\,N)} \frac{1}{(\bar{h}_1\bar{h}_2\cdots\bar{h}_s)^\alpha}. \tag{4.8}$$

Then for $f \in E_\alpha(c)$, we can write (4.5) as

$$|Q(\mathbf{z}, N)f - If| \leqslant cP_\alpha(\mathbf{z}, N), \tag{4.9}$$

with the bound being achieved if $f = cf_\alpha$. This motivates us to use $P_\alpha(\mathbf{z}, N)$ as a criterion for assessing rank-1 rules. Every such rule finds the function f_α a difficult one to integrate because the expression in (4.7) for the error in $Q(\mathbf{z}, N)f_\alpha$ involves no cancellation. Later we shall extend the definition of P_α to arbitrary lattice rules, and use P_α to compare rules of different rank.

In practice $P_\alpha(\mathbf{z}, N)$ is calculated as the quadrature error in f_α, since f_α is expressible as a product of functions of a single variable:

$$f_\alpha(\mathbf{x}) = \prod_{k=1}^{s} F_\alpha(x_k), \tag{4.10}$$

where

$$F_\alpha(x) = \sum_{h=-\infty}^{\infty} \frac{e^{2\pi\imath h x}}{\bar{h}^\alpha} = 1 + \sum_{h\neq 0} \frac{e^{2\pi\imath h x}}{|h|^\alpha}. \tag{4.11}$$

Normally α is taken to be an even integer, since then F_α can be expressed in terms of the Bernoulli polynomial B_α (see Appendix C),

$$B_\alpha(x) := -\alpha! \sum_{h\neq 0} \frac{e^{2\pi\imath h x}}{(2\pi\imath h)^\alpha}, \qquad x \in [0, 1], \qquad \alpha \geqslant 2,$$

the explicit relation for even α being

$$F_\alpha(x) = 1 - (-1)^{\frac{\alpha}{2}} \frac{(2\pi)^\alpha B_\alpha(x)}{\alpha!}, \qquad x \in [0, 1], \quad \alpha \text{ even}. \tag{4.12}$$

Thus for α an even integer, $P_\alpha(\mathbf{z}, N)$ is easily calculated by using

$$P_\alpha(\mathbf{z}, N) = Q(\mathbf{z}, N)f_\alpha - 1, \tag{4.13}$$

where f_α is given by (4.10) and (4.12).

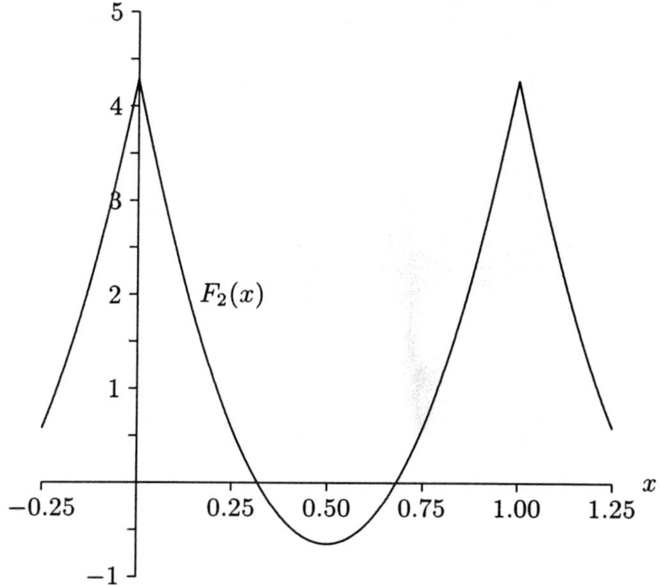

Fig. 4.1 $F_2(x) = 1 + 2\pi^2(x^2 - x + 1/6)$ on $[0, 1]$.

For the first three even values of α we have, explicitly, for $x \in [0, 1]$:

$$F_2(x) = 1 + 2\pi^2(x^2 - x + 1/6), \tag{4.14}$$

$$F_4(x) = 1 + \frac{\pi^4}{45}\left(1 - 30x^2(1 - x)^2\right), \tag{4.15}$$

$$F_6(x) = 1 + \frac{2\pi^6}{945}(1 - 21x^2 + 105x^4 - 126x^5 + 42x^6), \tag{4.16}$$

together with, in every case,

$$F_\alpha(x) = F_\alpha(x + 1), \quad x \in \mathbb{R}.$$

Graphs of those three functions are shown in Figures 4.1–4.3. Note that F_2 is continuous and has a simple discontinuity in its first derivative, F_4 has two continuous derivatives, and F_6 has four continuous derivatives. Another property, easily proved from (4.11), is that $F_\alpha(x)$ approaches $1 + 2\cos(2\pi x)$, uniformly in x, as $\alpha \to \infty$. It is obvious from Figure 4.3 that F_6 is a good approximation to this limit!

Tables of $\mathbf{z}$ obtained by minimizing $P_2(\mathbf{z}, N)$ may be found in Maisonneuve (1972) and Haber (1983). Other tables of $\mathbf{z}$ may be found in Saltykov (1963) (these are reproduced in Korobov (1963) and Stroud (1971)). The Saltykov tables, however, were obtained by minimizing a slightly different quantity, namely

$$\tilde{P}_2(\mathbf{z}, N) = Q(\mathbf{z}, N)\tilde{f}_2 - I\tilde{f}_2, \tag{4.17}$$

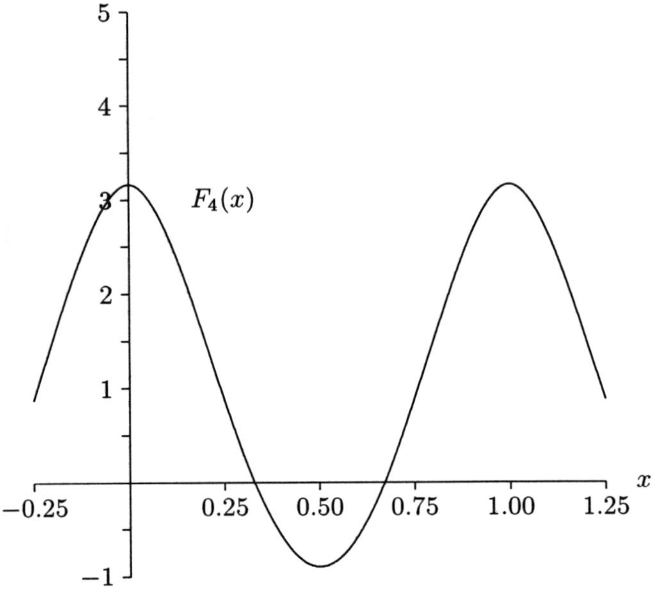

Fig. 4.2 $F_4(x) = 1 + \pi^4(1 - 30x^2(1 - x)^2)/45$ on $[0, 1]$.

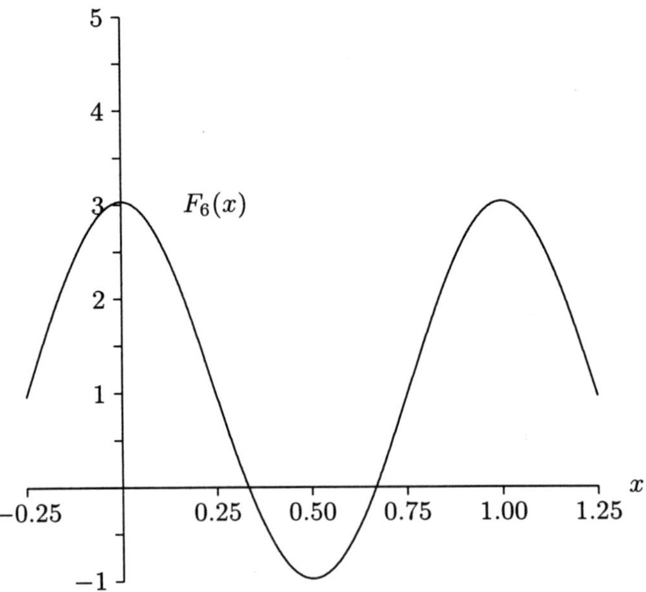

Fig. 4.3 $F_6(x) = 1 + 2\pi^6(1 - 21x^2 + 105x^4 - 126x^5 + 42x^6)/945$ on $[0, 1]$.

where

$$\tilde{f}_2(\mathbf{x}) = \prod_{k=1}^{s} \tilde{F}_2(x_k),$$

in which the function

$$\tilde{F}_2(x) = 1 + \frac{6}{\pi^2} \sum_{h \neq 0} \frac{e^{2\pi\imath hx}}{|h|^2} = 1 + \frac{6}{\pi^2}(F_2(x) - 1) = 3(2x - 1)^2$$

replaces $F_2(x)$. Perhaps the reason for using $\tilde{F}_2$ instead of F_2 was simply that π was an inconvenient number to work with in the early days of computers. Whatever the reason, the changed definition does make a difference, both to the 'best' values of $\mathbf{z}$ that are obtained in a search, and to the theoretical error estimate. We see that we can write $\tilde{P}_2(\mathbf{z}, N)$ in the form

$$\tilde{P}_2(\mathbf{z}, N) = \sideset{}{'}\sum_{\mathbf{h} \cdot \mathbf{z} \equiv 0 \,(\mathrm{mod}\, N)} \frac{1}{(\tilde{h}_1 \tilde{h}_2 \cdots \tilde{h}_s)^2}, \tag{4.18}$$

where $\tilde{h} = \max(1, \pi|h|/\sqrt{6})$. It follows, as remarked by Korobov (1960), that $P_2(\mathbf{z}, N)$ and $\tilde{P}_2(\mathbf{z}, N)$ satisfy, from (4.8) and (4.18),

$$\tilde{P}_2(\mathbf{z}, N) \leqslant P_2(\mathbf{z}, N) \leqslant \left(\frac{\pi^2}{6}\right)^s \tilde{P}_2(\mathbf{z}, N),$$

so that $P_2(\mathbf{z}, N)$ and $\tilde{P}_2(\mathbf{z}, N)$ can be used interchangeably up to constant factors. Unfortunately, the second constant $(\pi^2/6)^s$ is very much greater than 1 when s is large (for example, it is approximately 21 000 when $s = 20$), making the constant factors not insignificant. (Note that $\tilde{F}_2$ has smaller values of the higher Fourier coefficients than does F_2, and hence is smoother.) We shall always use the strict definition (4.8) for $P_\alpha(\mathbf{z}, N)$.

The largest terms in (4.7) are of the form $1/\rho(\mathbf{z}, N)^\alpha$, where

$$\rho(\mathbf{z}, N) = \rho(Q(\mathbf{z}, N)) := \min_{\substack{\mathbf{h} \cdot \mathbf{z} \equiv 0 \,(\mathrm{mod}\, N) \\ \mathbf{h} \neq \mathbf{0}}} \overline{h}_1 \overline{h}_2 \cdots \overline{h}_s. \tag{4.19}$$

This suggests that we can use $\rho(\mathbf{z}, N)$ as another possible measure of the goodness of a rank-1 rule: from this point of view, a good rank-1 rule is one in which $\rho(\mathbf{z}, N)$ is as large as possible. Apparently, $\rho(\mathbf{z}, N)$ was first introduced by Babenko (1960a,b). It is known as the figure of merit, and also as the Zaremba index (Lyness 1989).

It follows from the definitions of $P_\alpha(\mathbf{z}, N)$ and $\rho(\mathbf{z}, N)$ in (4.8) and (4.19) that

$$\frac{2}{\rho(\mathbf{z}, N)^\alpha} \leqslant \frac{\tau}{\rho(\mathbf{z}, N)^\alpha} \leqslant P_\alpha(\mathbf{z}, N), \tag{4.20}$$

where $\tau = \tau(Q(\mathbf{z}, N))$ is the number of integer vectors $\mathbf{h}$ satisfying $\mathbf{h} \cdot \mathbf{z} \equiv 0 \,(\mathrm{mod}\, N)$ and $\mathbf{h} \neq \mathbf{0}$ that achieve the minimum in (4.19). Note that $\tau \geqslant 2$, since if $\mathbf{h}$ satisfies $\mathbf{h} \cdot \mathbf{z} \equiv 0 \,(\text{mod } N)$ then so does $-\mathbf{h}$. An inequality in the opposite direction is obtained by Niederreiter (1978a), who shows that there exists a constant $d(s, \alpha)$, dependent only on s and α, for which

$$P_\alpha(\mathbf{z}, N) \leqslant d(s, \alpha) \frac{(\log N)^{s-1}}{\rho(\mathbf{z}, N)^\alpha}. \tag{4.21}$$

Taken together, these support the view given above that a good rank-1 rule should have as large a value of $\rho(\mathbf{z}, N)$ as possible.

Another criterion that can be used to assess lattice rules is the quantity $R(\mathbf{z}, N)$ (Niederreiter 1978a,b), defined by

$$R(\mathbf{z}, N) = R(Q(\mathbf{z}, N)) := \underset{\substack{\mathbf{h} \cdot \mathbf{z} \equiv 0 \,(\text{mod } N) \\ \mathbf{h} \in W(N)}}{\sum\nolimits'} \frac{1}{\overline{h_1}\,\overline{h_2} \cdots \overline{h_s}}, \tag{4.22}$$

where $W(N) = \{\mathbf{h} \in \mathbb{Z}^s : -\frac{N}{2} < h_k \leqslant \frac{N}{2}, 1 \leqslant k \leqslant s\}$. If we define

$$\phi(\mathbf{x}) := \sum_{\mathbf{h} \in W(N)} \frac{1}{\overline{h_1}\,\overline{h_2} \cdots \overline{h_s}} e^{2\pi i \mathbf{h} \cdot \mathbf{x}},$$

then it follows from (4.3) that $R(\mathbf{z}, N)$ can be written as the quadrature error

$$R(\mathbf{z}, N) = Q(\mathbf{z}, N)\phi - I\phi = Q(\mathbf{z}, N)\phi - 1. \tag{4.23}$$

Note that

$$\phi(\mathbf{x}) = \prod_{k=1}^{s} \Phi(x_k),$$

where

$$\Phi(x) = \sum_{-N/2 < h \leqslant N/2} \frac{e^{2\pi i h x}}{\overline{h}} = 1 + \sum_{-N/2 < h \leqslant N/2}^{\prime} \frac{e^{2\pi i h x}}{|h|}.$$

An asymptotic expansion for Φ can be used to provide an efficient technique for calculating $R(\mathbf{z}, N)$ by way of (4.23) (Joe and Sloan 1992b). A similar, but less accurate, approximation has also been used by Korobov (1963).

We see that $R(\mathbf{z}, N)$ is similar to $P_\alpha(\mathbf{z}, N)$ with $\alpha = 1$, except that the function ϕ replaces the function f_α. (Notice that $P_1(\mathbf{z}, N)$ is not defined, since the Fourier series for f_α is not absolutely convergent when $\alpha = 1$.) A relationship between $P_\alpha(\mathbf{z}, N)$, $\alpha > 1$, and $R(\mathbf{z}, N)$ is given (Niederreiter 1992b, Theorem 5.5) by

$$
\begin{aligned}
P_\alpha(\mathbf{z}, N) \;&<\; R(\mathbf{z}, N)^\alpha + (1 + 2\zeta(\alpha)N^{-\alpha})^s - 1 \\
&\quad + \frac{1}{N}(1 + 2\zeta(\alpha) + 2^\alpha\zeta(\alpha)N^{1-\alpha})^s - \frac{1}{N}(1 + 2\zeta(\alpha))^s \\
&=\; R(\mathbf{z}, N)^\alpha + O(N^{-\alpha}),
\end{aligned}
\tag{4.24}
$$

provided each component of $\mathbf{z}$ is relatively prime with N, where

$$
\zeta(\alpha) = \sum_{k=1}^{\infty} \frac{1}{k^\alpha}, \quad \alpha > 1,
$$

is the Riemann zeta function. The bound (4.24) strengthens earlier results (Niederreiter 1978a).

We conclude this section with a very brief discussion of another possible criterion, the 'discrepancy', which plays a role when the integrand f is of lesser smoothness than has been assumed so far. In particular, in the following discussion f is not required to have a continuous periodic extension, nor to have an absolutely convergent Fourier series. The following material is discussed with admirable completeness by Niederreiter (1978a).

Let J be the s-dimensional interval

$$
J = [u_1, v_1) \times \cdots \times [u_s, v_s),
$$

where

$$
0 \leqslant u_k < v_k \leqslant 1, \quad 1 \leqslant k \leqslant s,
$$

and let $|J|$ be the volume

$$
|J| = \prod_{k=1}^{s} (v_k - u_k).
$$

Then the discrepancy of a set $\{\mathbf{x}_0, \ldots, \mathbf{x}_{N-1}\}$ of points in U^s is defined by

$$
D_N := \sup_J \left| \frac{A(J)}{N} - |J| \right|,
\tag{4.25}
$$

where $A(J)$, the counting function, is the number of points of the set which lie in J. In effect the discrepancy is the supremum of the difference between $|J|$, which is the proportion of the points we would like to have in J, and the actual proportion $A(J)/N$, taken over all intervals J.

Its role in quadrature derives from the Koksma–Hlawka inequality (Hlawka 1961), a simple form of which states that

$$
\left| \frac{1}{N} \sum_{j=0}^{N-1} f(\mathbf{x}_j) - If \right| \leqslant V(f)D_N,
$$

where $V(f)$ is the variation of f in the sense of Hardy and Krause.

For the particular case in which the points $\mathbf{x}_0, \ldots, \mathbf{x}_{N-1}$ are the quadrature points of the rank-1 rule (4.1) we write $D_N = D(\mathbf{z}, N)$. For this case Niederreiter (1978a) obtains the following simple bound in terms of $R(\mathbf{z}, N)$:

$$D(\mathbf{z}, N) \leqslant \frac{s}{N} + \frac{1}{2} R(\mathbf{z}, N).$$

4.3   Fibonacci lattice rules

In this section we consider a very special class of two-dimensional rank-1 rules, the Fibonacci rules devised by Bahvalov (1959).

Recall that the Fibonacci numbers $1, 1, 2, 3, 5, 8, 13, \ldots$ are defined by

$$F_1 = F_2 = 1, \qquad F_k = F_{k-1} + F_{k-2} \text{ for } k \geqslant 3.$$

For each Fibonacci number F_k with $k \geqslant 3$ there is a corresponding Fibonacci lattice rule of order F_k

$$Q_k f = \frac{1}{F_k} \sum_{j=0}^{F_k - 1} f\left(\left\{ \frac{j}{F_k}(1, F_{k-1}) \right\} \right).$$

For example, on setting $k = 9$ we have a rule of order $N = F_9 = 34$:

$$Q_9 f = \frac{1}{34} \sum_{j=0}^{33} f\left(\left\{ \frac{j}{34}(1, 21) \right\} \right).$$

The quadrature points for this rule are shown in Figure 4.4.

The special quality of the Fibonacci lattice rules is seen most clearly in the figure of merit defined by (4.19): Zaremba (1966) showed that for the Fibonacci lattice rules this takes the simple value

$$\rho_k = \rho((1, F_{k-1}), F_k) = F_{k-2}, \quad k \geqslant 3. \tag{4.26}$$

It is easy to verify that ρ_k could not be any larger than F_{k-2}, since from the properties of the Fibonacci numbers we have

$$(F_{k-2}, 1) \cdot (1, F_{k-1}) = F_{k-2} + F_{k-1} = F_k \equiv 0 \,(\mathrm{mod}\, F_k),$$

so that $\pm(F_{k-2}, 1)$ are points of the dual lattice (4.2). The assertion in (4.26) is that there are no points $\mathbf{h}$ of the dual lattice closer to $\mathbf{0}$ than this, as measured by the values of $\overline{h}_1 \overline{h}_2$. The reader may wish to verify that in Figure 4.5, the dual of the lattice in Figure 4.4, there are no dual lattice points with $\overline{h}_1 \overline{h}_2$ smaller than 13.

Zaremba's proof of (4.26) is by means of a continued fraction argument. For those readers who know about such things, the key feature of the

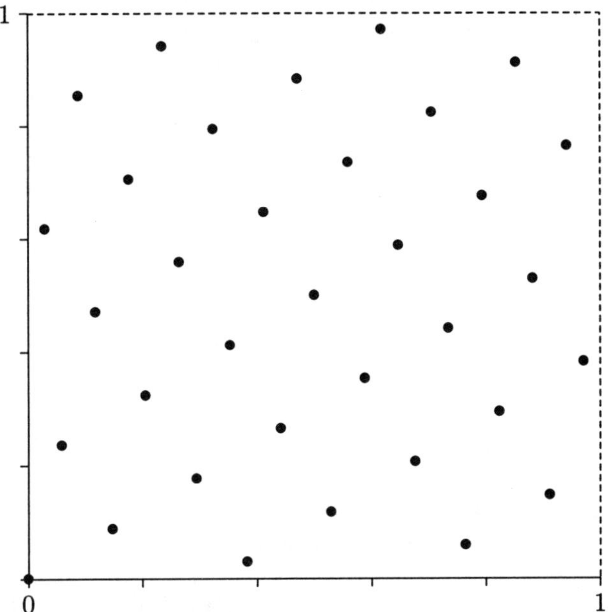

Fig. 4.4 The quadrature points for the Fibonacci lattice rule with $N = 34$.

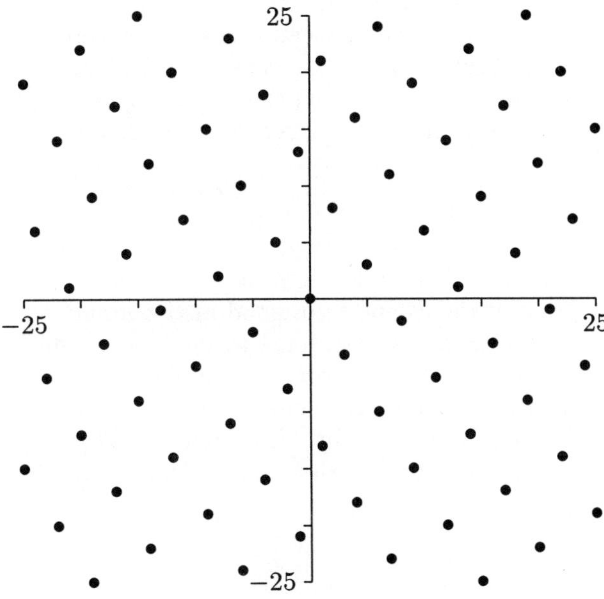

Fig. 4.5 Dual lattice for the $N = 34$ Fibonacci lattice.

Fibonacci numbers for this purpose is that the ratio $\frac{F_{k-1}}{F_k}$ of successive Fibonacci numbers has a continued fraction representation in which every partial quotient has the value 1. For a thorough discussion of this and other questions related to the Fibonacci lattice rules, see Niederreiter (1992*b*, Section 5.2).

Since $\frac{F_{k-2}}{F_k} \rightarrow \left(\frac{2}{1+\sqrt{5}}\right)^2$ as $k \rightarrow \infty$, we see from (4.26) that for the Fibonacci rules $\rho(\mathbf{z}, N) = O(N)$, so that the bound (4.21) gives

$$P_\alpha(\mathbf{z}, N) \leqslant d(\alpha)\frac{\log N}{N^\alpha}, \tag{4.27}$$

a result first obtained by Bahvalov (1959). This result sets a standard for all higher-dimensional lattice rules.

4.4 'Good' rank-1 rules

For the rank-1 rules given by (4.1), Korobov (1959) called the components of a vector $\mathbf{z}$ 'optimal coefficients' if the relationship

$$|Q(\mathbf{z}, N)f - If| \leqslant cd(s, \alpha)\frac{(\log N)^{\beta(s,\alpha)}}{N^\alpha}, \tag{4.28}$$

holds for functions $f \in E_\alpha(c)$, with $d(s, \alpha)$ and $\beta(s, \alpha)$ independent of N, and $\alpha > 1$. With a similar definition, Hlawka called such a vector $\mathbf{z}$ a 'good lattice point' (and so what we now call rank-1 rules came to be known as the method of good lattice points). Thus we shall use the abbreviation g.l.p. to denote a vector $\mathbf{z}$ (or, more properly, a sequence $\mathbf{z}(N)$) satisfying (4.28). It then follows from (4.9) that if a sequence of vectors $\mathbf{z}(N)$ satisfies

$$P_\alpha\left(\mathbf{z}(N), N\right) \leqslant d(s, \alpha)\frac{(\log N)^{\beta(s,\alpha)}}{N^\alpha} \tag{4.29}$$

for some constant $d(s, \alpha)$, then $\mathbf{z}$ is a g.l.p.

The first proof of the existence of good lattice points was that of Korobov (1959). That proof was restricted to the case in which N is prime. Subsequently, the existence was proved for simple composite numbers such as the product of two primes (Korobov 1960). Eventually the result was proved for all integers $N > 0$ by Niederreiter (1978*b*). In that work Niederreiter showed that there exists a $\mathbf{z} = (z_1, \ldots, z_s) \in \mathbb{Z}^s$ such that $\gcd(z_i, N) = 1$ for $1 \leqslant i \leqslant s$ and

$$R(\mathbf{z}, N) \leqslant \frac{1}{N}(2\log N + 1.4)^s, \tag{4.30}$$

from which there follows, with the aid of (4.24):

Theorem 4.2. *If $N \geqslant 2$ is any integer and $s \geqslant 2$, there exists a $\mathbf{z} \in \mathbb{Z}^s$ such that*

$$P_\alpha(\mathbf{z}, N) \leqslant \frac{(2 \log N)^{\alpha s}}{N^\alpha} + O\left(\frac{(\log N)^{\alpha s - 1}}{N^\alpha}\right). \tag{4.31}$$

In the next section we will show how a smaller bound can be obtained by a different route.

This bound, like the earlier one of Korobov (1959), has the exponent $\beta(s, \alpha)$ in (4.28) given by $\beta(s, \alpha) = \alpha s$. When N is prime Bahvalov (1959) shows that this can be reduced to $\beta(s, \alpha) = \alpha(s - 1)$. In the opposite direction, the work of Šarygin (1963) shows that $\beta(s, \alpha)$ is at least $s - 1$, with equality holding for $s = 2$. More recently, Niederreiter (1992c, 1993) has obtained bounds that are valid for composite N and have $\beta(s, \alpha)$ as either $\alpha(s - 1)$ or $\alpha(s - 1) + 1$. However, the given implied constants are dependent on N; in particular, on quantities relating to the details of the composite nature of N. Interested readers may consult these two references for further details.

Once a fixed value of $\alpha > 1$ has been chosen, for given values of N and s one can carry out a computer search to find the vector $\mathbf{z}$ that minimizes $P_\alpha(\mathbf{z}, N)$, secure in the knowledge that this minimizer must be a g.l.p. Because of (4.1), the components of $\mathbf{z}$ may be replaced by their residues modulo N without changing $Q(\mathbf{z}, N)f$, and so we need consider at most N^s possible choices of $\mathbf{z}$. However, for large values of N and s, N^s is much too large a number to search through. For this reason it is common to limit the search to vectors $\mathbf{z}$ of a restricted form, the most common of which is the one-parameter form used by Korobov (1960) in his existence proofs, namely

$$\mathbf{z}(\ell) = (1, \ell, \ell^2 \bmod N, \ldots, \ell^{s-1} \bmod N), \quad 1 \leqslant \ell < N. \tag{4.32}$$

Some of the $\mathbf{z}$ vectors in the tables of Saltykov (1963), Maisonneuve (1972), and Haber (1983) are of such a form. With this choice we need only search through $N - 1$ possible choices of $\mathbf{z}$, instead of N^s. Indeed, we can achieve even greater efficiency: by using (4.10) and the fact that

$$F_\alpha(x) = F_\alpha(1 - x), \quad x \in [0, 1],$$

it is easily seen that $f_\alpha(\{j\mathbf{z}(\ell)/N\}) = f_\alpha(\{j\mathbf{z}(N - \ell)/N\})$, so that the work can be halved by searching over $\mathbf{z}(\ell)$ for $1 \leqslant \ell \leqslant \lfloor N/2 \rfloor$ rather than $1 \leqslant \ell < N$. Similarly, in a full search it is sufficient to restrict each component of $\mathbf{z}$ to the interval $0 \leqslant z_k \leqslant \lfloor \frac{N}{2} \rfloor$, so reducing the number of z values in the search by a factor of approximately 2^s. In Section 10.6, we shall see that it is sometimes possible to reduce the computational work even more by omitting some 'geometrically equivalent' rules.

The Korobov form is a good choice on theoretical as well as practical grounds (Korobov 1960, 1963, Hua and Wang 1981). Essentially the result presented in these works is that if N is a prime number, then there is an integer $\ell(N)$ such that $\mathbf{z}(\ell(N))$ is a g.l.p. Though the theoretical results for the restricted Korobov form of $\mathbf{z}$ have been obtained only for N prime, the numerical evidence (Haber 1983) indicates that such a form also works well for general N. Presumably, for general N the techniques in Niederreiter (1978b) could be used to obtain a theoretical result for the Korobov form. For $s = 3$, such a result may be found in Larcher and Niederreiter (1989).

Other special forms of $\mathbf{z}$ which can be used to obtain a g.l.p. are also known. For instance, when N is a product of two distinct primes, a suitable form may be found in Korobov (1960), and a more general case when N is a product of more than two distinct primes may be found in Hua and Wang (1981). Additionally, Hua and Wang (1981) proposed another method for obtaining good rank-1 rules based on ideas from algebraic number theory and rational approximation. It is quite efficient for finding suitable vectors $\mathbf{z}$, especially for large values of s, but for fixed N and s, the values of $P_\alpha(\mathbf{z}, N)$ obtained from such vectors are usually not as good as those obtained from searching with the Korobov form $\mathbf{z}(\ell)$ given in (4.32); moreover, this approach can only be applied to certain sequences of N values.

As pointed out by Haber (1983), 'good lattice points' are not unique or rare, but are quite common (we shall see why this is so in the next section). Thus he proposed and tested a search procedure (Haber 1972, 1983) by which, for given N and s, a fixed number of $\mathbf{z}$ vectors were generated pseudo-randomly, with the best one, in the sense of minimizing $P_\alpha(\mathbf{z}, N)$, being selected for use in (4.1). The numerical evidence was encouraging.

4.5   Improved bounds on $P_\alpha(\mathbf{z}, N)$

According to Theorem 4.2 in the preceding section, there exists $d(s, \alpha)$ and there exist integer vectors $\mathbf{z} = \mathbf{z}(N)$ such that for large N

$$P_\alpha(\mathbf{z}, N) \leqslant d(s, \alpha) \frac{(\log N)^{\alpha s}}{N^\alpha}.$$

It is convenient to introduce an asymptotic bound with an explicit constant,

$$P_\alpha(\mathbf{z}, N) \lesssim 2^{\alpha s} \frac{(\log N)^{\alpha s}}{N^\alpha}, \tag{4.33}$$

with the symbol $\lesssim$ denoting that the quantity on the right is the dominant error term when N is large.

An asymptotic bound with a smaller constant factor has been obtained by Disney and Sloan (1991) and Disney (1990): following a different line of argument, they obtain the following result.

Theorem 4.3. *If $s \geqslant 3$, there exists a $\mathbf{z} \in \mathbb{Z}^s$ such that*

$$P_\alpha(\mathbf{z}, N) \lesssim \left(\frac{2e}{s}\right)^{\alpha s} \frac{(\log N)^{\alpha s}}{N^\alpha}$$

as $N \to \infty$.

Note that the constant in this asymptotic bound decreases faster than exponentially for large s, and is less than 1 for $s \geqslant 6$, whereas that in (4.33) increases rapidly with s. The ratio $(s/e)^{\alpha s}$ of the constants can be truly astronomical when s is large; for example it has the value 4.7×10^{34} when $s = 20$ and $\alpha = 2$.

To date, the best bounds on $P_\alpha(\mathbf{z}, N)$ have been those obtained by Niederreiter (1992a,c, 1993). The coefficients of the leading terms in these bounds decrease at about the same rate as the $(2e/s)^{\alpha s}$ factor here. However, these bounds are more complicated than the bound given in Theorem 4.3 and their proofs are rather technical. For these reasons, we shall not consider them here.

If N is prime, Disney and Sloan (1991) obtain the following explicit bound. Most of our effort in this section will be devoted to proving this result.

Theorem 4.4. *If $\alpha > 1$, N is prime, $N > e^{\alpha s/(\alpha-1)}$, and $s \geqslant 2$, then there exists a $\mathbf{z} \in \mathbb{Z}^s$ such that*

$$P_\alpha(\mathbf{z}, N) \leqslant \left(\frac{e}{s}\right)^{\alpha s} \frac{(2\log N + s)^{\alpha s}}{N^\alpha}. \tag{4.34}$$

The basic technique used by Disney and Sloan (1991) and Disney (1990) to obtain their results is to derive an expression for the mean of $P_\alpha(\mathbf{z}, N)$ over a certain subset of vectors $\mathbf{z}$. An analogous expression for the mean of $R(\mathbf{z}, N)$ over the same vectors was derived by Niederreiter (1978b) to obtain the bound (4.30).

Definition 4.5. *For any integer $N \geqslant 2$, let $Z = Z(N)$ be the set of all $\mathbf{z} \in \mathbb{Z}^s$ whose components z_k are relatively prime to N and satisfy $-N/2 < z_k \leqslant N/2$. The mean of $P_\alpha(\mathbf{z}, N)$ over $\mathbf{z} \in Z$ is*

$$M_\alpha(N) := \frac{1}{\varphi(N)^s} \sum_{\mathbf{z} \in Z} P_\alpha(\mathbf{z}, N), \quad \alpha > 1. \tag{4.35}$$

Here φ is Euler's totient function (that is, $\varphi(N)$ is the number of positive integers less than N and relatively prime to N); of course $\varphi(N)^s$ is simply the number of terms in the sum in (4.35). In the case when N is prime, $\varphi(N)$ is simply $N - 1$. According to the definition, the mean is taken over all vectors $\mathbf{z}$ which have components relatively prime to N and which satisfy $-N/2 < z_k \leqslant N/2$. Since the components of $\mathbf{z}$ may be replaced by their residues modulo N without changing $Q(\mathbf{z}, N)f$, it makes sense to allow each z_k to range only over a convenient set of representatives for the residue class modulo N.

The importance of the mean $M_\alpha(N)$ lies in this simple fact: that because it is an average of $P_\alpha(\mathbf{z}, N)$ over a set, there must always exist at least one $\mathbf{z}$ in the set for which $P_\alpha(\mathbf{z}, N)$ is no larger than the mean. This obvious result may be formalized in the following manner.

Proposition 4.6. *Let $N \geqslant 2$ be any integer and let Z be the set defined in Definition 4.5. Then there exists a $\mathbf{z} \in Z$ such that*

$$P_\alpha(\mathbf{z}, N) \leqslant M_\alpha(N).$$

This result may seem rather feeble. Nevertheless, it is a valuable tool for proving the existence of good choices of $\mathbf{z}$, since any bound on $M_\alpha(N)$ immediately gives us an existence result. We shall obtain such bounds (by way of an explicit expression for $M_\alpha(N)$) below.

However, before we turn to computing the mean, it is useful, following Disney and Sloan (1991), to strengthen Proposition 4.6 in the following way.

Proposition 4.7. *Let $N \geqslant 2$ be any integer, let Z be the set defined in Definition 4.5, and let γ satisfy $1 < \gamma < \alpha$. Then there exists a $\mathbf{z} \in Z$ such that*

$$P_\alpha(\mathbf{z}, N) \leqslant M_\gamma(N)^{\alpha/\gamma}.$$

Proof Jensen's inequality (Hardy *et al.* 1934, Theorem 19) states that if $\{a_k\}$ is a sequence of positive numbers, then

$$\left(\sum a_k^p \right)^{1/p} \leqslant \left(\sum a_k^q \right)^{1/q}, \text{ if } 0 < q < p.$$

From the definition of $P_\alpha(\mathbf{z}, N)$ in (4.8), we see that this implies

$$P_\alpha(\mathbf{z}, N) \leqslant P_\gamma(\mathbf{z}, N)^{\alpha/\gamma}, \text{ if } 1 < \gamma < \alpha.$$

Since there is at least one $\mathbf{z}$ such that $P_\gamma(\mathbf{z}, N)$ is no larger than $M_\gamma(N)$, the result follows immediately.   ■

This result can be used in just the same way as Proposition 4.6, in that every bound for the mean automatically gives us an existence result. The difference is that we may choose γ in any way that suits us best. Later we shall exploit this freedom in the choice of γ.

We now give an explicit expression for $M_\alpha(N)$, due to Disney and Sloan (1991), for the case when N is prime.

Theorem 4.8. *If N is a prime number, then*

$$M_\alpha(N) = \frac{(1 + 2\zeta(\alpha))^s}{N} + \frac{N-1}{N} \left(1 - \frac{2(1 - N^{1-\alpha})\zeta(\alpha)}{N-1} \right)^s - 1. \quad (4.36)$$

Proof Using (4.1) and (4.13), we can write $P_\alpha(\mathbf{z}, N)$ as

$$P_\alpha(\mathbf{z}, N) = \frac{1}{N} f_\alpha(\mathbf{0}) + \left[\frac{1}{N} \sum_{j=1}^{N-1} f_\alpha\left(\left\{\frac{j}{N}\mathbf{z}\right\}\right) \right] - 1,$$

in which the $j = 0$ term in the quadrature sum has been separated out because it is independent of $\mathbf{z}$. Now from (4.10) and (4.11) we have

$$f_\alpha(\mathbf{0}) = \prod_{k=1}^{s}\left(1 + \sum_{h\neq 0} \frac{1}{|h|^\alpha} \right) = (1 + 2\zeta(\alpha))^s, \tag{4.37}$$

and therefore

$$P_\alpha(\mathbf{z}, N) = \frac{(1 + 2\zeta(\alpha))^s}{N} + \left[\frac{1}{N} \sum_{j=1}^{N-1} f_\alpha\left(\left\{\frac{j}{N}\mathbf{z}\right\}\right) \right] - 1.$$

Since N is prime, $\varphi(N) = N - 1$. Thus from the definition of $M_\alpha(N)$ in (4.35), we have

$$
\begin{aligned}
M_\alpha(N) &= \frac{1}{(N-1)^s} \sum_{\mathbf{z}\in Z} \left[\frac{(1 + 2\zeta(\alpha))^s}{N} + \left[\frac{1}{N} \sum_{j=1}^{N-1} f_\alpha\left(\left\{\frac{j}{N}\mathbf{z}\right\}\right) \right] - 1 \right] \\
&= \frac{(1 + 2\zeta(\alpha))^s}{N} + \left[\frac{1}{N(N-1)^s} \sum_{\mathbf{z}\in Z} \sum_{j=1}^{N-1} f_\alpha\left(\left\{\frac{j}{N}\mathbf{z}\right\}\right) \right] - 1,
\end{aligned}
$$

in which the means of the first and third terms, which are independent of $\mathbf{z}$, are obtained trivially. Writing this as

$$M_\alpha(N) = \frac{(1 + 2\zeta(\alpha))^s}{N} + \theta_\alpha(N) - 1,$$

we then obtain from (4.6)

$$
\begin{aligned}
\theta_\alpha(N) &= \frac{1}{N(N-1)^s} \sum_{j=1}^{N-1} \sum_{\mathbf{z}\in Z} \sum_{\mathbf{h}\in\mathbb{Z}^s} \frac{1}{(\overline{h}_1 \overline{h}_2 \cdots \overline{h}_s)^\alpha} e^{2\pi\imath j\mathbf{h}\cdot\mathbf{z}/N} \\
&= \frac{1}{N(N-1)^s} \sum_{j=1}^{N-1} \underset{-N/2<z_1\leqslant N/2}{\sum{}'} \cdots \underset{-N/2<z_s\leqslant N/2}{\sum{}'} \\
&\qquad \sum_{h_1=-\infty}^{\infty} \cdots \sum_{h_s=-\infty}^{\infty} \frac{e^{2\pi\imath j h_1 z_1/N} \cdots e^{2\pi\imath j h_s z_s/N}}{(\overline{h}_1 \overline{h}_2 \cdots \overline{h}_s)^\alpha}
\end{aligned}
$$

$$= \frac{1}{N} \sum_{j=1}^{N-1} \left(\frac{1}{N-1} \sideset{}{'}\sum_{-N/2<z\leqslant N/2} \sum_{h=-\infty}^{\infty} \frac{e^{2\pi i jhz/N}}{\overline{h}^{\alpha}} \right)^{s}.$$

Thus, we can write

$$\theta_{\alpha}(N) = \frac{1}{N} \sum_{j=1}^{N-1} \left(\frac{1}{N-1} T_{\alpha}(j,N) \right)^{s}, \tag{4.38}$$

where

$$T_{\alpha}(j,N) = \sum_{h=-\infty}^{\infty} \sideset{}{'}\sum_{-N/2<z\leqslant N/2} \frac{e^{2\pi i jhz/N}}{\overline{h}^{\alpha}}, \quad 1 \leqslant j \leqslant N-1.$$

Now, separating out the terms in which h is a multiple of N, we obtain

$$\frac{1}{N-1} T_{\alpha}(j,N) \;=\; \frac{1}{N-1} \left(\sum_{k=-\infty}^{\infty} \sideset{}{'}\sum_{-N/2<z\leqslant N/2} \frac{1}{(\overline{Nk})^{\alpha}} \right.$$

$$\left. + \sum_{h\not\equiv 0\,(\mathrm{mod}\,N)} \frac{1}{|h|^{\alpha}} \sideset{}{'}\sum_{-N/2<z\leqslant N/2} e^{2\pi i jhz/N} \right)$$

$$=\; 1 + \frac{2\zeta(\alpha)}{N^{\alpha}}$$

$$+ \frac{1}{N-1} \sum_{h\not\equiv 0\,(\mathrm{mod}\,N)} \frac{1}{|h|^{\alpha}} \sideset{}{'}\sum_{-N/2<z\leqslant N/2} e^{2\pi i jhz/N}.$$

But for $h \not\equiv 0\,(\mathrm{mod}\,N)$ and $1 \leqslant j \leqslant N-1$ we have

$$\sideset{}{'}\sum_{-N/2<z\leqslant N/2} e^{2\pi i jhz/N} = \sum_{-N/2<z\leqslant N/2} (e^{2\pi i jh/N})^{z} - 1 = -1,$$

since jh is not a multiple of N for j in this range and N prime. (This is
the only point in the proof where the assumption that N is prime matters.)
Thus

$$\frac{1}{N-1} T_{\alpha}(j,N) \;=\; 1 + \frac{2\zeta(\alpha)}{N^{\alpha}} + \frac{1}{N-1} \sum_{h\not\equiv 0\,(\mathrm{mod}\,N)} \frac{-1}{|h|^{\alpha}}$$

$$=\; 1 + \frac{2\zeta(\alpha)}{N^{\alpha}} - \frac{1}{N-1} \left(\sum_{h\neq 0} \frac{1}{|h|^{\alpha}} - \sum_{k\neq 0} \frac{1}{|Nk|^{\alpha}} \right)$$

$$=\; 1 + \frac{2\zeta(\alpha)}{N^{\alpha}} - \frac{2}{N-1} \left(\zeta(\alpha) - \frac{\zeta(\alpha)}{N^{\alpha}} \right)$$

$$=\; 1 - \frac{2(1 - N^{1-\alpha})\zeta(\alpha)}{N-1}.$$

Since $T_{\alpha}(j,N)$ has turned out to be independent of j, substitution of this last expression into (4.38) yields

$$\theta_\alpha(N) = \frac{N-1}{N}\left(1 - \frac{2(1 - N^{1-\alpha})\zeta(\alpha)}{N-1}\right)^s,$$

and hence the desired result. ∎

We see from the proof of Theorem 4.8 that the term $(1 + 2\zeta(\alpha))^s/N$ in (4.36) is just $f_\alpha(0)/N$, which is always present in $P_\alpha(\mathbf{z}, N)$ irrespective of the value of $\mathbf{z} \in Z$. It turns out that this term is a major contributor to $M_\alpha(N)$ when N is large. In fact for $N \geqslant \zeta(\alpha) + 1$ we have

$$1 > 1 - \frac{2(1 - N^{1-\alpha})\zeta(\alpha)}{N-1} > 1 - \frac{2\zeta(\alpha)}{N-1} \geqslant -1,$$

which leads to the following corollary.

Corollary Suppose N is a prime number satisfying $N \geqslant \zeta(\alpha) + 1$. Then

$$M_\alpha(N) \leqslant \frac{(1 + 2\zeta(\alpha))^s}{N}.$$

Using Proposition 4.6 as well as the previous theorem and its corollary, we obtain the following theorem on the existence of good $\mathbf{z}$.

Theorem 4.9. *If N is a prime number, then there exists a $\mathbf{z} \in Z$ such that*

$$P_\alpha(\mathbf{z}, N) \leqslant \frac{(1 + 2\zeta(\alpha))^s}{N} + \frac{N-1}{N}\left(1 - \frac{2(1 - N^{1-\alpha})\zeta(\alpha)}{N-1}\right)^s - 1.$$

If $N \geqslant \zeta(\alpha) + 1$ then for this choice of $\mathbf{z}$

$$P_\alpha(\mathbf{z}, N) \leqslant \frac{(1 + 2\zeta(\alpha))^s}{N}.$$

The bounds in Theorem 4.9 are only of order N^{-1}, and therefore cannot be very good for large values of N. Nevertheless, Disney and Sloan (1991) present persuasive numerical evidence that the bounds are numerically reasonable for practical values of N and s, and, except for extraordinary large values of N, are often the best available.

To obtain the bound of order $O((\log N)^{\alpha s}/N^\alpha)$ stated in Theorem 4.4 we may proceed by way of Proposition 4.7.

Proof of Theorem 4.4 It follows from Proposition 4.7 and the corollary to Theorem 4.8 that there exists a $\mathbf{z} \in Z$ such that

$$P_\alpha(\mathbf{z}, N) \leqslant \left(\frac{(1 + 2\zeta(\gamma))^s}{N}\right)^{\alpha/\gamma}, \tag{4.39}$$

where γ is any number satisfying $1 < \gamma < \alpha$ and provided N satisfies $N \geqslant \zeta(\gamma) + 1$. The result of Theorem 4.4 comes from making a special choice of γ, namely,

$$\gamma = \frac{\log N}{\log N - s}.$$

With this choice it is easy to see that $1 < \gamma < \alpha$ is satisfied when

$$N > e^{\alpha s/(\alpha-1)}.$$

Moreover, on using

$$\zeta(\gamma) = \sum_{k=1}^{\infty} \frac{1}{k^\gamma} < 1 + \int_1^\infty \frac{1}{x^\gamma}\, dx = 1 + \frac{1}{\gamma - 1} = \frac{\log N}{s}, \qquad (4.40)$$

we see that $\log N > s\zeta(\gamma)$ and hence the requirement that $N \geqslant \zeta(\gamma) + 1$ is automatically satisfied. Now we have

$$\frac{1}{\gamma} = 1 - \frac{s}{\log N},$$

which leads to

$$N^{-\alpha/\gamma} = N^{-\alpha} e^{\alpha s}.$$

Also from (4.40) we have

$$1 + 2\zeta(\gamma) < \frac{2\log N + s}{s},$$

and thus we see that (4.39) yields

$$\begin{aligned}
P_\alpha(\mathbf{z}, N) \ &\leqslant\ \frac{(1 + 2\zeta(\gamma))^{\alpha s}}{N^{\alpha/\gamma}} \\
&<\ \frac{(2\log N + s)^{\alpha s}}{s^{\alpha s}} \frac{e^{\alpha s}}{N^\alpha} = \left(\frac{e}{s}\right)^{\alpha s} \frac{(2\log N + s)^{\alpha s}}{N^\alpha},
\end{aligned}$$

which completes the proof. ∎

Better bounds are obtained by Disney and Sloan (1991) by a similar argument, but with a sharper bound for $M_\gamma(N)$, and with a more carefully chosen value of γ. For fuller details the reader is referred to the original paper.

We conclude this section with an explicit expression for $M_\alpha(N)$ for general N (Disney 1990), but omit the proof as it relies on number-theoretic arguments that are beyond the scope of this book.

Theorem 4.10. *If $N \geqslant 2$ is any integer, then*

$$M_\alpha(N) = \frac{1}{N} \sum_{k=1}^{s} \binom{s}{k} (2\zeta(\alpha))^k \prod_{p|N} G_{\alpha,k}(p^\epsilon), \qquad (4.41)$$

where the product is over all primes p which divide N, p^ϵ is the highest *power of p which divides N, and*

$$G_{\alpha,k}(p^\epsilon) = 1 + (-1)^k \frac{(1 - 1/p^{\alpha-1})^k (1 - 1/p^{(\alpha k-1)\epsilon})}{(p-1)^{k-1}(1 - 1/p^{\alpha k-1})}.$$

This theorem and Proposition 4.7 with a well-chosen value of γ can be used to derive the asymptotic bound in Theorem 4.3; we shall omit the details. When N is prime, the result of Theorem 4.10 reduces to that of Theorem 4.8. (Hint: use the binomial theorem.) As can be seen, Theorem 4.10 in the general case is rather involved, with the expression for $M_\alpha(N)$ in (4.41) depending on the prime factorization of N. By obtaining bounds on $\prod_{p|N} G_{\alpha,k}(p^\epsilon)$, Disney (1990) was able to replace the identity in (4.41) with a bound on $M_\alpha(N)$ which is much simpler to use.

Corollary

$$M_\alpha(N) \leqslant \frac{1}{N}[a(1 + 2\zeta(\alpha))^s + b(1 - 2\zeta(\alpha))^s] + \frac{2s(s-1)\zeta(\alpha)^s}{\varphi(N)},$$

where $a = \zeta(3)/\zeta(6) + 0.5 = 1.68\ldots$ and $b = a - 1$.

4.6 Error estimation by randomization

In general, it is not possible to obtain useful error estimates for rank-1 rules by repeating the calculation with an increasing number of quadrature points, since the true errors tend to fluctuate erratically. The reason for this is that the quadrature points for different rank-1 rules usually have nothing in common, having been obtained by independent numerical searches. Expressed in another way, suppose we have two rank-1 rules with orders N and N' and with generating vectors $\mathbf{z}$ and $\mathbf{z}'$ respectively. Then the corresponding dual lattices given by $\{\mathbf{h} \in \mathbb{Z}^s : \mathbf{h} \cdot \mathbf{z} \equiv 0 \,(\mathrm{mod}\,N)\}$ and $\{\mathbf{h} \in \mathbb{Z}^s : \mathbf{h} \cdot \mathbf{z}' \equiv 0 \,(\mathrm{mod}\,N')\}$ usually bear no relation to each other.

A quite different strategy, proposed by Cranley and Patterson (1976), is to make use of a fixed rank-1 rule (a 'good' one, of course!), but to employ it in the shifted form (2.21):

$$Q(\mathbf{z}, N, \mathbf{c})f = \frac{1}{N} \sum_{j=0}^{N-1} f\left(\left\{\frac{j}{N}\mathbf{z} + \mathbf{c}\right\}\right), \tag{4.42}$$

with $\mathbf{c}$ in this application being a random vector. By making several calculations with different randomly chosen $\mathbf{c}$, it is possible to obtain not only an estimate of the integral, but also confidence intervals for the magnitude of the error. Discussion of this randomization procedure requires a few basic concepts from statistical theory. A text such as Lindgren (1976) may be useful to readers not familiar with these ideas.

We recall from (2.22) that the error for the shifted rank-1 rule $Q(\mathbf{z}, N, \mathbf{c})$ is given by

$$Q(\mathbf{z}, N, \mathbf{c})f - If = {\sum_{\mathbf{h}\cdot\mathbf{z}\equiv 0\,(\mathrm{mod}\,N)}}^{\prime} e^{2\pi i \mathbf{h}\cdot\mathbf{c}}\,\hat{f}(\mathbf{h}). \qquad (4.43)$$

We see from (4.3) and (4.43) that the errors for $Q(\mathbf{z}, N)$ and $Q(\mathbf{z}, N, \mathbf{c})$ depend on the same set of Fourier coefficients, and that there is no reason to believe that any one member of the family of shifted rank-1 rules would be any better for a general integrand f than any other. In the terminology of Cranley and Patterson (1976), $Q(\mathbf{z}, N, \mathbf{c})$ is a stochastic family of quadrature rules.

Suppose that the random vector $\mathbf{c}$ is chosen from a multivariate uniform distribution on C^s so that each component of $\mathbf{c}$ is independently uniformly distributed on $[0, 1]$. It turns out that the stochastic family of shifted rank-1 rules given in (4.42) is an unbiased estimator of If; that is, the centre of gravity of the distribution of $Q(\mathbf{z}, N, \mathbf{c})f$ is If.

Theorem 4.11. *Suppose $\mathbf{c}$ has a multivariate uniform distribution on C^s. Then*

$$E(Q(\mathbf{z}, N, \mathbf{c})f) = If,$$

where $E(\cdot)$ is the expectation operator.

Proof It follows from (4.43) that we can obtain the desired result by proving that

$$E\left({\sum_{\mathbf{h}\cdot\mathbf{z}\equiv 0\,(\mathrm{mod}\,N)}}^{\prime} \hat{f}(\mathbf{h})e^{2\pi i \mathbf{h}\cdot\mathbf{c}}\right) = 0.$$

Since the expectation operator is linear, the expectation on the left is a linear combination of terms of the form

$$E\left(e^{2\pi i \mathbf{h}\cdot\mathbf{c}}\right) = \int_{C^s} e^{2\pi i \mathbf{h}\cdot\mathbf{c}}\,d\mathbf{c} = \prod_{k=1}^{s}\int_{0}^{1} e^{2\pi i h_k c_k}\,dc_k = 0 \text{ for } \mathbf{h} \neq \mathbf{0},$$

which completes the proof. ∎

An immediate consequence of Theorem 4.11 is the following corollary.

Corollary For any positive integer q, let $\mathbf{c}_1, \ldots, \mathbf{c}_q$ be independent random vectors with a multivariate uniform distribution on C^s. Then the estimate

$$\bar{Q}(\mathbf{z}, N)f = \frac{1}{q}\sum_{k=1}^{q} Q(\mathbf{z}, N, \mathbf{c}_k)f$$

is an unbiased estimate of If.

Thus we see that an unbiased approximation to If may be obtained by taking the mean of q shifted rank-1 rules. Calculation of this estimate requires Nq function evaluations. The standard error of this approximation is given by $\sigma = \varsigma/\sqrt{q}$, where ς^2 is the variance of $Q(\mathbf{z}, N, \mathbf{c})f$. It follows from the well-known Chebyshev inequality (Lindgren 1976, p.132) that

$$\text{Probability}(|\bar{Q}(\mathbf{z}, N)f - If| < v\sigma) \geqslant 1 - \frac{1}{v^2}, \quad v > 1,$$

so that it is possible to calculate a confidence interval for the error. Suppose we calculate, for the same value of q, many values of $\bar{Q}(\mathbf{z}, N)f$, and at the same time the corresponding intervals $(\bar{Q}(\mathbf{z}, N)f - v\sigma, \bar{Q}(\mathbf{z}, N)f + v\sigma)$ for a given choice of v. Then, loosely speaking, the Chebyshev inequality tells us that the proportion of these intervals that contain the true value of If is at least $1 - 1/v^2$. Alternatively, we could say that we had at least $(100 - 100/v^2)\%$ 'confidence' that $v\sigma$ was an upper bound on the true error.

In general we do not know what the true value of σ is. Thus we would normally use the estimate of σ given by

$$\tilde{\sigma} = \sqrt{\frac{\displaystyle\sum_{k=1}^{q}\left(Q(\mathbf{z}, N, \mathbf{c}_k)f - \bar{Q}(\mathbf{z}, N)f\right)^2}{q(q-1)}}$$

to obtain the desired confidence interval. This estimate of σ is the one used in the NAG implementation of the rank-1 method (see the documentation for the routines D01GCF and D01GDF in Numerical Algorithms Group 1991). There it is recommended that q be chosen to be in the range 3–5.

5

Lattice rules of higher rank—a first look

5.1 Introduction

In this chapter we begin the quantitative study of lattice rules of higher rank. In Chapters 2 and 3 we looked at the structure of lattice rules and at their representations as multiple sums. Now we turn to the question of how good such rules might be.

The first step is to decide on a criterion for measuring the goodness of a lattice rule. In Section 5.2 the quantities P_α, ρ, and R are defined for a general lattice, by extending the definitions already introduced for rank-1 rules in the preceding chapter. As in the rank-1 case, our own choice will be the quantity P_α, because it is the quadrature error for a particular function which every lattice rule finds especially difficult. It is also a bound on the quadrature error for an important class of functions.

Do good lattice rules of higher rank exist? The rank-1 rules discussed in the preceding chapter set the standard. We have already established the existence of rank-1 rules which have impressive rates of convergence with respect to the total number of quadrature points N. Can rules of higher rank achieve the same rates of convergence? We begin our quantitative study, in Section 5.3, with a simple demonstration that lattice rules of any rank r, and with any fixed set of higher invariants $n_2, \ldots, n_r$ can achieve exactly the same order of convergence as the best rank-1 rule.

By itself this result does not tell us very much, because the order of convergence is not the only thing that matters: just as important is the associated constant factor. A moment of thought will persuade us that two methods with the same order of convergence may perform very differently in practice, if they are allowed to have different constant factors in front. (As a matter of fact the proof mentioned above produces remarkably bad constants for the rules of higher rank. However, that is an artifact of the proof: in effect the proof merely argues that a rank-r rule with $n_1 n_2 \cdots n_r$ points is as good as a certain rank-1 rule with just n_1/n_2 points, obtained by throwing out most of the original points.) This is a theme that will recur: if we want to compare two different methods, we must be prepared to consider not only the powers of N that express the order of convergence, but also the constant factors that stand before them.

In Section 5.3 we also describe a result for the particular case of rank-2 rules obtained recently by Niederreiter (1992*a*). This is a better result than the general result referred to above, in that it is shown that a rank-2 rule with invariants n_1, n_2 can achieve the same order of convergence as the best rank-1 rule, even if the second invariant n_2 grows slowly (whereas in the result referred to above n_2 must be held constant). However, since the argument keeps no control over the constant factors, it, too, provides no basis for a quantitative comparison between rank-2 and rank-1 rules.

The first evidence that rules of higher rank might be quantitatively competitive with the traditional rules of rank 1 came from numerical experiments (Sloan and Walsh 1990) for rules of rank 2. These experiments are described briefly in Section 5.4. At the time those results were obtained no theoretical predictions of any kind were available. Since that time a much improved theoretical understanding of the relation between rank-1 rules and rules of higher rank has emerged. This will be the subject of Chapters 6 and 7.

5.2 'Goodness' criteria for general lattice rules

We now define quantities $P_\alpha(L)$, $\rho(L)$, $R(L)$ that generalize in an obvious way the definitions given in Chapter 4.

Let $Q(L)$ denote the lattice rule that corresponds to the integration lattice L. Beginning with $P_\alpha(L)$, we define, for $\alpha > 1$,

$$P_\alpha(L) = P_\alpha(Q(L)) := {\sum_{\mathbf{h} \in L^\perp}}' \frac{1}{(\overline{h}_1 \overline{h}_2 \cdots \overline{h}_s)^\alpha}. \tag{5.1}$$

From the fundamental error expression (2.20), we recognize that $P_\alpha(L)$ is the error in the lattice rule $Q(L)$ applied to a certain function: specifically,

$$P_\alpha(L) = Q(L)f_\alpha - 1, \tag{5.2}$$

where, for completeness, we recall from (4.10) and (4.11) that

$$f_\alpha(\mathbf{x}) = \prod_{k=1}^{s} F_\alpha(x_k),$$

with

$$F_\alpha(x) = \sum_{h=-\infty}^{\infty} \frac{e^{2\pi i h x}}{\overline{h}^\alpha} = 1 + \sum_{h \neq 0} \frac{e^{2\pi i h x}}{|h|^\alpha}.$$

For $f \in E_\alpha(c)$ we have the rigorous error bound

$$|Q(L)f - If| \leqslant cP_\alpha(L),$$

which follows easily from (2.20) and the definition of $E_\alpha(c)$. It is the property (5.2) that makes $P_\alpha(L)$ so easy to calculate.

In a similar way, we may define, analogously to (4.19) and (4.22),

$$\rho(L) = \rho(Q(L)) := \min_{\substack{\mathbf{h} \in L^\perp \\ \mathbf{h} \neq \mathbf{0}}} \overline{h}_1 \overline{h}_2 \cdots \overline{h}_s, \tag{5.3}$$

and

$$R(L) = R(Q(L)) := \sideset{}{'}\sum_{\substack{\mathbf{h} \in L^\perp \\ \mathbf{h} \in W(N)}} \frac{1}{\overline{h}_1 \overline{h}_2 \cdots \overline{h}_s}.$$

Finally, the discrepancy $D(L) = D(Q(L))$ can be defined by (4.25), just as it can be for any point set in U^s.

Various inequalities between the quantities $P_\alpha(L)$, $\rho(L)$, $R(L)$, and $D(L)$ are known for general lattice rules. First, from Sloan and Kachoyan (1987) there exists $d(s, \alpha)$ such that

$$P_\alpha(L) \leqslant \frac{d(s, \alpha)}{\rho(L)^\alpha}(1 + \log \rho(L))^{s-1},$$

while in the other direction an elementary inequality analogous to (4.20) holds for any lattice rule, that is,

$$\frac{2}{\rho(L)^\alpha} \leqslant \frac{\tau(L)}{\rho(L)^\alpha} \leqslant P_\alpha(L),$$

where $\tau(L) = \tau(Q(L))$ is the number of integer vectors $\mathbf{h} \in L^\perp$ which achieve the minimum in (5.3). Note that $\tau(L) \geqslant 2$, since if $\mathbf{h} \in L^\perp$ then so is $-\mathbf{h}$. The next two inequalities are from Niederreiter and Sloan (1990):

$$R(L) \leqslant \frac{1}{\rho(L)} \left(\frac{2}{\log 2} \right)^{s-1} \left(2(\log N)^s + 3(\log N)^{s-1} \right), \quad N \geqslant 5,$$

$$\frac{1}{d(s)\rho(L)} \leqslant D(L) \leqslant \frac{s}{N} + \frac{1}{2}R(L),$$

where $d(s) = 2((\pi + 1)^s - 1)/\pi$ for $s \geqslant 4$, $d(3) = 27$, and $d(2) = 4$. Finally, Niederreiter (1992a) shows that a result similar to (4.24) holds for any lattice rule $Q(L)$, namely,

$$P_\alpha(L) \leqslant (1 + 2\zeta(\alpha))^s R(L)^\alpha + O(N^{-\alpha}). \tag{5.4}$$

For the case $\alpha \geqslant 2$, this bound has been improved (Niederreiter 1992b, Theorem 5.26) to

$$P_\alpha(L) \leqslant (1 + \zeta(\alpha)s)R(L)^\alpha + O(N^{-\alpha}).$$

The inequalities in the preceding paragraph hold for any lattice rule $Q(L)$. Niederreiter (1992b) points out that better inequalities can be obtained if the invariants $n_1, \ldots, n_r$ are known. Two elementary inequalities of this kind are

$$\rho(L) \leqslant n_1,$$

and

$$D(L) \geqslant 1 - \left(1 - \frac{1}{n_1}\right)^s \geqslant \frac{1}{n_1}.$$

The first holds because $L^\perp$ contains $(n_1, 0, \ldots, 0)$. To see this, merely observe that n_1 is divisible by each of the smaller invariants $n_2, \ldots, n_r$, so that $(n_1, 0, \ldots, 0) \cdot (j_1 \mathbf{z}_1/n_1 + \cdots + j_r \mathbf{z}_r/n_r) \in \mathbb{Z}$.

While each of the quantities $P_\alpha(L)$, $\rho(L)$, $R(L)$, and $D(L)$ (and no doubt others, too) might be used as a measure of the lattice rule $Q(L)$, our favoured candidate throughout will be $P_\alpha(L)$, with the parameter $\alpha > 1$ chosen for convenience. Remember that this is the quadrature error for the function f_α, which is the most difficult function for any lattice rule to integrate out of the whole class $E_\alpha(1)$. Because it is a quadrature error, it is very easy to calculate, which is not the case for the other quantities.

The following very simple theorem is sometimes useful. It says that if new points are added to the lattice, $P_\alpha(L)$ can only decrease.

Theorem 5.1. *If the integration lattice L has a proper sublattice L' which is also an integration lattice, then*

$$P_\alpha(L) < P_\alpha(L').$$

Proof We are given $L' \subset L$, from which it follows on using the definition (2.19) of the dual lattice that $L^\perp \subset (L')^\perp$. Now

$$P_\alpha(L') = \sideset{}{'}\sum_{\mathbf{h} \in (L')^\perp} \frac{1}{(\overline{h}_1 \overline{h}_2 \cdots \overline{h}_s)^\alpha},$$

in which each term is positive. Since $P_\alpha(L)$ is obtained from this merely by omitting some of these positive terms, the result follows immediately. ∎

5.3 Do there exist 'good' rules of higher rank?

First we recall some necessary definitions from Chapter 3. A lattice rule of rank r, with $1 \leqslant r \leqslant s$, has uniquely defined invariants $n_1, \ldots, n_r$, which satisfy

$$n_{k+1} \text{ divides } n_k \text{ for } 1 \leqslant k \leqslant r - 1.$$

Such a rule can be expressed in the form

$$Q(L)f = \frac{1}{N} \sum_{j_r=0}^{n_r-1} \cdots \sum_{j_1=0}^{n_1-1} f\left(\left\{\frac{j_1}{n_1}\mathbf{z}_1 + \cdots + \frac{j_r}{n_r}\mathbf{z}_r\right\}\right), \tag{5.5}$$

where $\mathbf{z}_1, \ldots, \mathbf{z}_r$ are integer vectors. The number of quadrature points is

$$N = n_1 n_2 \cdots n_r,$$

and the lattice L that corresponds to the rule (5.5) is

$$L = \left\{\left\{\frac{j_1}{n_1}\mathbf{z}_1 + \cdots + \frac{j_r}{n_r}\mathbf{z}_r\right\} : \ 0 \leqslant j_k \leqslant n_k - 1, \ 1 \leqslant k \leqslant r\right\} + \mathbb{Z}^s.$$

In the preceding chapter we have seen that rank-1 rules can perform very well indeed, in that $P_\alpha(\mathbf{z}, N)$ can approach zero very rapidly as N, the number of quadrature points, approaches ∞. Our question here is whether higher-rank rules can perform as well as the best of the rank-1 rules. A first simple answer is that they certainly can if the only test is the order of convergence of $P_\alpha(L)$ as N approaches ∞.

Specifically, the lattice rule (5.5) with rank r and invariants $n_1, \ldots, n_r$ has $n_1 n_2 \cdots n_r$ quadrature points. Setting $n_1/n_2 = n$, consider the subset of these points obtained by setting $j_1 = jn_2$ for $0 \leqslant j \leqslant n - 1$ and $j_2 = \cdots = j_r = 0$ in (5.5). These are the quadrature points of a rank-1 lattice rule

$$Q(L')f = \frac{1}{n} \sum_{j=0}^{n-1} f\left(\left\{\frac{j}{n}\mathbf{z}_1\right\}\right),$$

whose lattice L' is given by

$$L' = \left\{\left\{\frac{j}{n}\mathbf{z}_1\right\} : \ 0 \leqslant j \leqslant n - 1\right\} + \mathbb{Z}^s,$$

which is of course a subset of L.

Now because $Q(L')$ is a rank-1 lattice rule, we know from the results reported in Chapter 4 that $P_\alpha(L')$ can be made small by an appropriate choice of $\mathbf{z}_1$. To be specific, we choose to make use of the best result known, which is that due to Bahvalov (1959): it says that if n is prime there exists $\mathbf{z}_1$ such that

$$P_\alpha(L') = P_\alpha(\mathbf{z}_1, n) \leqslant d(s, \alpha) \frac{(\log n)^{\alpha(s-1)}}{n^\alpha}, \tag{5.6}$$

where $d(s, \alpha)$ is independent of n. It then follows from Theorem 5.1 that

$$P_\alpha(L) \leqslant P_\alpha(L') \leqslant d(s, \alpha) \left(\frac{N}{n}\right)^\alpha \frac{(\log N)^{\alpha(s-1)}}{N^\alpha},$$

since $n \leqslant N$. Thus if the higher invariants $n_2, \ldots, n_r$ are held constant, so that N/n is constant, then $P_\alpha(L)$ achieves the same order of convergence as that found by Bahvalov for rules of rank 1.

Summarizing, we have the following trivial theorem.

Theorem 5.2. *Let* $r \geqslant 1$ *and* $n_2, \ldots, n_r$ *be given integers such that* n_{k+1} *divides* n_k *for* $2 \leqslant k \leqslant r-1$ *and* $n_r \geqslant 2$. *Then for each prime number* n *there exists a lattice rule of rank* r *and invariants* $n_1 = nn_2, n_2, \ldots, n_r$ *such that*

$$P_\alpha(L) \leqslant e(s, \alpha) \frac{(\log N)^{\alpha(s-1)}}{N^\alpha},$$

where $N = n_1 n_2 \cdots n_r$ *and* $e(s, \alpha)$ *is independent of* n.

An analogous result which holds for all values of n, whether prime or not, but with the exponent of $\log N$ becoming αs instead of $\alpha(s-1)$, is obtained by replacing Bahvalov's bound (5.6) by Niederreiter's bound (4.31).

Recently, a similar but deeper result than that stated above has been obtained by Niederreiter (1992a) for the particular case of rank-2 rules. He shows that for any invariants n_1 and n_2 there exists a rule $Q(L)$ of rank 2 such that

$$R(L) \leqslant \frac{d(s)}{N} \left((\log N)^s + n_2 \log N \right),$$

from which it follows, via (5.4), that

$$P_\alpha(L) \leqslant \frac{e(s, \alpha)}{N^\alpha} \left((\log N)^s + n_2 \log N \right)^\alpha, \text{ for all } \alpha > 1. \tag{5.7}$$

This result has the attractive feature that even if n_2 grows like $(\log N)^{s-1}$, we still achieve a bound of order $O((\log N)^{\alpha s}/N^\alpha)$ for $P_\alpha(L)$. On the other hand, the constants are still not controlled. Not until the next chapter will we see comparisons that take account of the constant factors.

5.4　Numerical experiments for the rank-2 case

To find out experimentally how good rank-2 rules might be, Sloan and Walsh (1990) carried out extensive numerical experiments, with the object of finding the best possible values of $P_\alpha(L)$. To make the competition fair, they studied both rank-1 and rank-2 rules, spending a comparable amount of effort on each. As we shall see, they found that some rank-2 rules performed marginally better than the best of the rank-1 rules.

In a rank-2 rule there are of course two invariants. Remembering that the second invariant must always divide the first, it is convenient to write the two invariants as nm and n, where $n \geqslant 2$ and $m \geqslant 1$. In the present setting the smaller invariant n is a fixed small number, such as 2, or 3, or 4.

Sloan and Walsh restricted attention to the case in which n and m are relatively prime. The reason for considering just this case is that, as we may recall from Section 3.6, in this situation the rank-2 rule can be written in the three-sum form

$$Q(L)f = \frac{1}{n^2 m} \sum_{k_2=0}^{n-1} \sum_{k_1=0}^{n-1} \sum_{j=0}^{m-1} f\left(\left\{\frac{j}{m}\mathbf{z} + \frac{k_1}{n}\mathbf{y}_1 + \frac{k_2}{n}\mathbf{y}_2\right\}\right), \qquad (5.8)$$

in which there is one more sum than is strictly necessary.

The advantage of this form over the ordinary two-sum canonical form is that the symmetry of the expression makes it easier to eliminate 'geometrically equivalent' rules from the search. We recall from Section 2.13 that two rules are geometrically equivalent if one can be obtained from the other by a relabelling of the coordinates, or by a reflection in one of the mid-planes perpendicular to the coordinate axes, or by a combination of such transformations. Since $P_\alpha(L)$ is unchanged under any such transformation, two geometrically equivalent rules will give exactly the same value of $P_\alpha(L)$, making it wasteful to calculate $P_\alpha(L)$ for more than one member of a geometrically equivalent family. This is particularly important if a systematic search is envisaged, since there may be an enormous number of rules which are distinct but nevertheless geometrically equivalent.

Sloan and Walsh tackled the problem of eliminating geometrical equivalence in the following simple way. Because n and m are relatively prime, the rule (5.8) can be written as a direct sum (see Definition 3.23 and Theorem 3.27)

$$Q(L) = q(\mathbf{z}) \oplus Q(l), \qquad (5.9)$$

where the first component is an m-point rank-1 rule

$$q(\mathbf{z})f = \frac{1}{m} \sum_{j=0}^{m-1} f\left(\left\{\frac{j}{m}\mathbf{z}\right\}\right),$$

and the second is a rank-2 rule with invariants n, n:

$$Q(l)f = \frac{1}{n^2} \sum_{k_2=0}^{n-1} \sum_{k_1=0}^{n-1} f\left(\left\{\frac{k_1}{n}\mathbf{y}_1 + \frac{k_2}{n}\mathbf{y}_2\right\}\right), \qquad (5.10)$$

with l denoting the corresponding integration lattice

$$l = \left\{\left\{\frac{k_1}{n}\mathbf{y}_1 + \frac{k_2}{n}\mathbf{y}_2\right\} : 0 \leqslant k_1, k_2 \leqslant n-1\right\} + \mathbb{Z}^s.$$

Now suppose that $Q(L)$ is geometrically equivalent to another $n^2 m$-point rule $Q(L')$. Then there is some linear operation T mapping the unit cube to itself such that $L' = TL$, from which it follows that

$$Q(L') = Q(TL) = q(T\mathbf{z}) \oplus Q(Tl).$$

Since we are prepared to treat $Q(L)$ and $Q(L') = Q(TL)$ as equivalent, it follows that we never need to consider more than one of $Q(l)$ and $Q(Tl)$; but to be sure that we do not strike out some truly inequivalent rules, we will then need to let $\mathbf{z}$ in (5.9) range freely. Since $Q(l)$ and $Q(Tl)$ are geometrically equivalent, we have now reduced the problem of identifying and containing geometrical equivalence to one involving just the n^2-point rules of the form (5.10).

Before we go further, it is time to face another problem that complicates a computer search of lattice rules: namely, the difficulty that a list of apparently different lattice rules may actually contain the same rule (let alone ones that are geometrically equivalent) many times over. This is because two rules with different choices of the integer vectors in (5.8) may turn out to be identical, there being no general theorem that the representation (5.8) is unique. To avoid such redundancies Sloan and Walsh restricted attention to a subclass of the rules with invariants nm and n, namely, those which are 'projection regular'. (From Definition 3.38, a rank-2 rule is projection regular if the projections on to the x_1 axis and the x_1–x_2 plane have as many points as possible, namely nm and n^2m respectively.) The advantage of the projection-regular rules is that in this case the rule (5.8) can be written in a form that is unique: Theorem 3.52 states that we may then choose

$$z_1 = 1, \qquad y_{11} = 1, \qquad y_{12} = 0, \qquad y_{21} = 0, \qquad y_{22} = 1 \qquad (5.11)$$

(where y_{kp} is the pth element of $\mathbf{y}_k$), and also

$$0 \leqslant z_p < m, \quad 2 \leqslant p \leqslant s, \qquad 0 \leqslant y_{kp} < n \quad \text{for } k = 1, 2, \quad 3 \leqslant p \leqslant s; \tag{5.12}$$

and that if we make this choice then all components of $\mathbf{z}$, $\mathbf{y}_1$, and $\mathbf{y}_2$ are uniquely determined.

The first step of the search operation for a given value of s and n was to construct a full list of vectors $\mathbf{y}_1$ and $\mathbf{y}_2$ which satisfied the constraints in the preceding paragraph and such that no two pairs in the list gave geometrically equivalent n^2-point rules. In practice, this was achieved by running through all vectors $\mathbf{y}_1$ and $\mathbf{y}_2$ satisfying the constraints in (5.11) and (5.12), for each such pair evaluating $Q(l)g$, where g is the function

$$g(\mathbf{x}) = \left(\sum_{k=1}^{s} G(x_k) \right)^s,$$

with

$$G(x) = \left| x - \frac{1}{2} \right|,$$

and then omitting from the list any pair $\mathbf{y}_1$, $\mathbf{y}_2$ giving a value seen already. (For a minor modification that reduces the calculation to one involving

Table 5.1 Geometrically inequivalent $\mathbf{y}_1$, $\mathbf{y}_2$ pairs for $s = 4$

n	$\mathbf{y}_1$				$\mathbf{y}_2$			
2	1	0	0	0	0	1	0	0
	1	0	0	0	0	1	0	1
	1	0	0	1	0	1	0	1
	1	0	0	0	0	1	1	1
	1	0	0	1	0	1	1	0
	1	0	0	1	0	1	1	1
3	1	0	0	0	0	1	0	0
	1	0	0	0	0	1	0	1
	1	0	0	1	0	1	0	1
	1	0	0	0	0	1	1	1
	1	0	0	1	0	1	1	0
	1	0	0	1	0	1	1	1
	1	0	1	1	0	1	1	2

only integers rather than floating point numbers, see the original reference.) Because this particular function is symmetric under all the operations that define geometric equivalence, it is clear that two geometrically equivalent rules $Q(l)$ and $Q(l')$ will give identical approximations to If. In the other direction, Sloan and Walsh conjecture, but do not prove, that geometrically inequivalent rules will always give different approximations to If. (If this turns out to be wrong then the lists of pairs of vectors $\mathbf{y}_1$, $\mathbf{y}_2$ obtained in this way will be incomplete.)

A list of all the geometrically inequivalent pairs for four dimensions and $n = 2$ and $n = 3$ is given in Table 5.1. Similar tables for other dimensions and other values of n are easily constructed. Once these lists of interestingly different pairs $\mathbf{y}_1$, $\mathbf{y}_2$ are available, the basic search procedure is easily stated: for each selected value of m (and hence of $N = n^2m$), one runs over all vectors $\mathbf{z}$ that satisfy the constraints in (5.11) and (5.12), as well as over all the vector pairs $\mathbf{y}_1$ and $\mathbf{y}_2$ in the list, and records the rule that gives the smallest values of $P_\alpha(L)$. When m is large the procedure may have to be modified by restricting attention to $\mathbf{z}$ vectors of the Korobov form $(1, \ell, \ell^2 \bmod m, \ldots, \ell^{s-1} \bmod m)$, or by selecting the components of $\mathbf{z}$ from the allowable set by using a pseudo-random number generator. Sloan and Walsh tried both techniques, generally finding the former the more effective for a given amount of search effort.

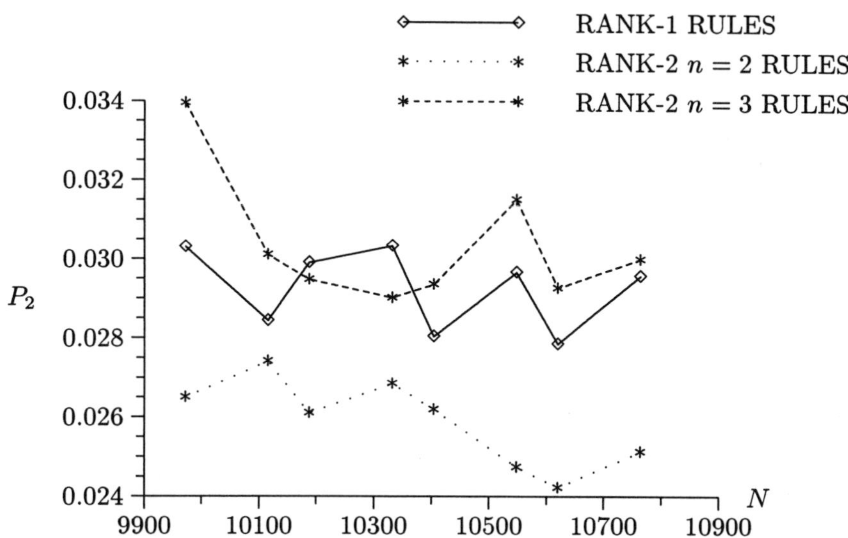

Fig. 5.1 P_2 as a function of N in $s = 5$ dimensions for the 10 000-point window.

Sloan and Walsh carried out searches in three 'windows' of N values, centred around 1000, 10 000 and 100 000 points. The purpose of looking at a window of values rather than just at single values of N is that in this way one may reduce the accidental effect of a single value of N that turns out to be particularly lucky or unlucky. They looked at values of the dimension s ranging from 3 to 8.

For a full description of the search procedure and the results obtained from it the reader is referred to the original paper. In brief, after a preliminary study Sloan and Walsh settled on the two smallest values $n = 2$ and $n = 3$ of the second invariant to study in detail, having established that larger values of n gave worse results. Of these two choices it was the smaller value that turned out almost invariably to give the better result.

Stated broadly, the best of the rank-2 results with smaller invariant $n = 2$ almost invariably gave smaller values of $P_\alpha(L)$ than the best of the rank-1 rules, with the picture becoming clearer as the dimensionality increases. Typical results are shown in Figures 5.1 and 5.2, for the case $\alpha = 2$, with $s = 5$ in the first case, and $s = 8$ in the second. (The lines joining the points in the graphs are included merely to guide the eye, and have no other significance.) The quantitative differences are not very large, but the fact that the pattern is so consistent is interesting at least from a theoretical point of view, suggesting as it does a theorem waiting to be discovered. Steps in this direction are discussed in Chapter 7.

One property discovered in these searches and deserving particular mention is this: that for the important $n = 2$ case the best rules turned out

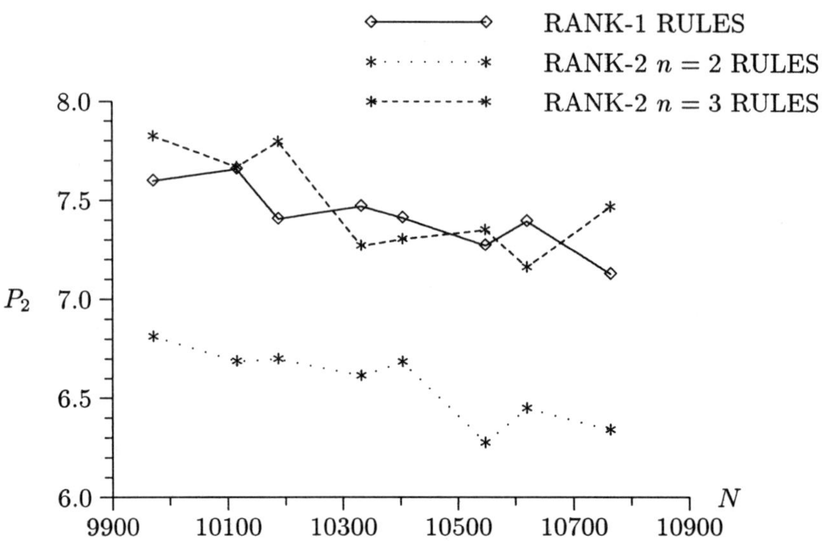

Fig. 5.2 P_2 as a function of N in $s = 8$ dimensions for the 10 000-point window.

almost invariably to arise from a particular $\mathbf{y}_1$, $\mathbf{y}_2$ pair. That favoured pair was the simplest of all, namely, $\mathbf{y}_1 = (1, 0, 0, \ldots, 0)$ and $\mathbf{y}_2 = (0, 1, 0, \ldots, 0)$. While not claiming to understand this bias, Sloan and Walsh exploited it effectively in their searches by restricting attention to just this case when the number of points was large. Theoretically the observation seems surprising, because from one point of view the favoured $\mathbf{y}_1$, $\mathbf{y}_2$ pair is the worst possible choice! For if one looks at just one-dimensional projections, the projections on to the x_1 axis are the same for every $\mathbf{y}_1$, $\mathbf{y}_2$ pair, as are the projections on to the x_2 axis, because of the constraints in (5.11), while for the remaining one-dimensional projections the favoured $\mathbf{y}_1, \mathbf{y}_2$ pair contributes nothing at all; whereas every other choice of $\mathbf{y}_1$, $\mathbf{y}_2$ contributes something to at least one of the one-dimensional projections. Obviously the one-dimensional projections are not the whole story.

While this experimentally observed preference for the pair $\mathbf{y}_1 = (1, 0, 0, \ldots, 0)$, $\mathbf{y}_2 = (0, 1, 0, \ldots, 0)$ is still not fully understood, substantial theoretical progress has now been made, since, as we shall see in Chapter 7, this simple choice of $\mathbf{y}_1$, $\mathbf{y}_2$ has now been shown to be 'good' in a certain precise sense. (Chapter 7 also shows that the values of $P_\alpha(L)$ for the n^2m-point rule (5.8) with this special choice of $\mathbf{y}_1$, $\mathbf{y}_2$ can be computed very efficiently, as a quadrature error for a rule with just m points, instead of n^2m points as in formula (5.2).)

6

Maximal rank lattice rules

6.1 Introduction

In this chapter we look at lattice rules that have the maximal rank s. We meet a simple characterization of maximal rank rules, and show that some maximal rank rules yield outstandingly attractive error bounds. Later in this book we shall see that maximal rank rules have other attributes that make them attractive in computational algorithms.

Our simple characterization of maximal rank rules is in terms of 'n^s copies' of lattice rules of lower rank. The n^s copy of any quadrature rule defined over C^s is obtained by subdividing the unit cube into n^s smaller cubes each of side $\frac{1}{n}$, and then applying an appropriately scaled version of the rule to each smaller cube. We shall see (in Section 6.4) that there is a compelling theoretical case for choosing $n = 2$. In the last section of this chapter we try to explain what it is about 2^s copies that makes them special.

For a lattice rule Q with integration lattice L, the n^s copy of Q is the lattice rule with lattice $n^{-1}L$. Thus we have the following definition.

Definition 6.1. *Suppose Q is the N-point lattice rule given by*

$$Qf = \frac{1}{N} \sum_{j=0}^{N-1} f(\mathbf{x}_j).$$

Then the n^s copy of Q is given by

$$\frac{1}{n^s N} \sum_{k_s=0}^{n-1} \cdots \sum_{k_1=0}^{n-1} \sum_{j=0}^{N-1} f\left(\frac{(k_1, k_2, \ldots, k_s)}{n} + \frac{\mathbf{x}_j}{n}\right).$$

Example 6.2. *If Q is the one-point lattice rule*

$$Qf = f(\mathbf{0}),$$

with lattice $L = \mathbb{Z}^s$, then the n^s copy of this lattice rule is the product-rectangle rule given by

$$\frac{1}{n^s} \sum_{k_s=0}^{n-1} \cdots \sum_{k_1=0}^{n-1} f\left(\frac{k_1}{n}, \ldots, \frac{k_s}{n}\right).$$

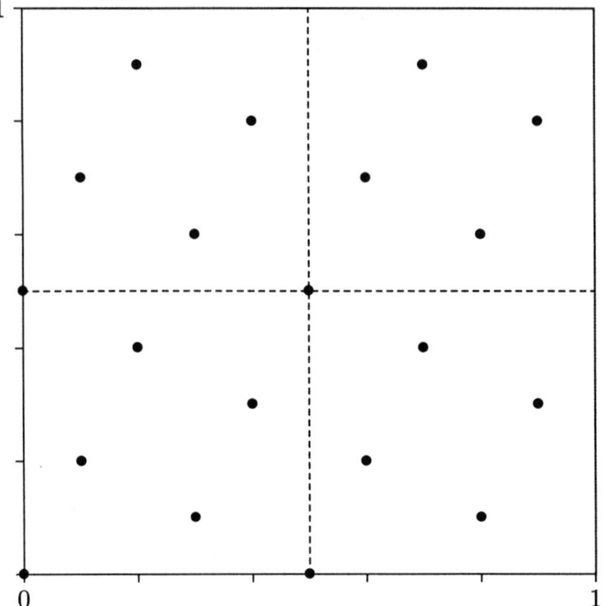

Fig. 6.1 The 2^2 copy of the rule in Figure 2.1.

This rule has the lattice $(1/n)\mathbb{Z}^s$. Moreover, it is clear from Theorem 3.2 that the rank of this rule is s, with the invariants being $n, n, \ldots, n$ (the n occurs s times).

Example 6.3. *The 2^2 copy of the two-dimensional five-point rule*

$$Qf = \frac{1}{5} \sum_{j=0}^{4} f\left(\left\{\frac{j}{5}(1,2)\right\}\right),$$

whose points are shown in Figure 2.1, is

$$\frac{1}{20} \sum_{k_2=0}^{1} \sum_{k_1=0}^{1} \sum_{j=0}^{4} f\left(\left\{\frac{(k_1,k_2)}{2} + \frac{j}{10}(1,2)\right\}\right).$$

The points of this 20-point rule are shown in Figure 6.1.

Notice that in writing down the n^s copy of a lattice rule we may reposition the 'fractional-part' braces as we have done in the last example, by appealing to the obvious identity

$$\frac{k}{n} + \frac{\{x\}}{n} = \frac{k - \lfloor x \rfloor}{n} + \frac{x}{n}.$$

6.2 Characterization of maximal rank rules

The following theorem, first proved by Sloan and Lyness (1989), gives our desired characterization of maximal rank rules.

Theorem 6.4. *An s-dimensional lattice rule is of rank s if and only if it is the n^s copy, with $n > 1$, of a rule of lower rank. More precisely, a rule with invariants $n_1, \ldots, n_s$, where $n_s > 1$, is the n_s^s copy of a rule of rank t and invariants $n_1/n_s, \ldots, n_t/n_s$, where $t < s$ is the largest integer for which $n_t > n_s$.*

Proof Suppose Q is a lattice rule of rank s with invariants $n_1, \ldots, n_s$. Then Theorem 3.2 shows that Q may be written in canonical form as

$$Qf = \frac{1}{n_1 n_2 \cdots n_s} \sum_{j_s=0}^{n_s-1} \cdots \sum_{j_1=0}^{n_1-1} f\left(\left\{\frac{j_1}{n_1}\mathbf{z}_1 + \cdots + \frac{j_s}{n_s}\mathbf{z}_s\right\}\right). \tag{6.1}$$

Since n_{i+1} divides n_i for $1 \leqslant i \leqslant s-1$, the invariant $n = n_s$ is a common factor of all the other invariants. So in (6.1) we may decompose the sum over j_i by setting $h_i = n_i/n$ and

$$j_i = k_i h_i + m_i, \quad 1 \leqslant i \leqslant s,$$

and then summing k_i over $0, 1, \ldots, n-1$ and m_i over $0, 1, \ldots, h_i - 1$. Thus we obtain

$$\begin{aligned} Qf \ = \ & \frac{1}{n^s h_1 h_2 \cdots h_s} \sum_{k_s=0}^{n-1} \cdots \sum_{k_1=0}^{n-1} \sum_{m_s=0}^{h_s-1} \cdots \sum_{m_1=0}^{h_1-1} \\ & \times f\left(\left\{\frac{k_1}{n}\mathbf{z}_1 + \cdots + \frac{k_s}{n}\mathbf{z}_s + \frac{m_1}{nh_1}\mathbf{z}_1 + \cdots + \frac{m_s}{nh_s}\mathbf{z}_s\right\}\right). \end{aligned} \tag{6.2}$$

Because Q has $n_1 n_2 \cdots n_s$ distinct quadrature points, each term in (6.2) must correspond to a distinct quadrature point. In particular, the n^s points

$$\left\{\frac{k_1}{n}\mathbf{z}_1 + \cdots + \frac{k_s}{n}\mathbf{z}_s\right\}, \quad 0 \leqslant k_i \leqslant n-1, \quad 1 \leqslant i \leqslant s,$$

must be distinct. It then follows that they must be just the quadrature points for the n^s-point product-rectangle rule taken in a different order. Thus we may write Q as

$$Qf \;=\; \frac{1}{n^s h_1 h_2 \cdots h_s} \sum_{k_s=0}^{n-1} \cdots \sum_{k_1=0}^{n-1} \sum_{m_s=0}^{h_s-1} \cdots \sum_{m_1=0}^{h_1-1}$$

$$\times f\left(\left\{ \frac{(k_1, k_2, \ldots, k_s)}{n} + \frac{m_1}{nh_1}\mathbf{z}_1 + \cdots + \frac{m_s}{nh_s}\mathbf{z}_s \right\}\right),$$

which from Definition 6.1 is just the n^s copy of the rule

$$\frac{1}{h_1 h_2 \cdots h_s} \sum_{m_s=0}^{h_s-1} \cdots \sum_{m_1=0}^{h_1-1} f\left(\left\{ \frac{m_1}{h_1}\mathbf{z}_1 + \cdots + \frac{m_s}{h_s}\mathbf{z}_s \right\}\right).$$

If $t < s$ is the largest integer for which $n_t > n_s$, then $h_{t+1} = \cdots = h_s = 1$, so this rule may be written in the canonical form

$$\frac{1}{h_1 h_2 \cdots h_t} \sum_{m_t=0}^{h_t-1} \cdots \sum_{m_1=0}^{h_1-1} f\left(\left\{ \frac{m_1}{h_1}\mathbf{z}_1 + \cdots + \frac{m_t}{h_t}\mathbf{z}_t \right\}\right).$$

Thus we see that Q is the n_s^s copy of a lattice rule of rank t having the invariants $h_1, \ldots, h_t$, that is, having the invariants $n_1/n_s, \ldots, n_t/n_s$.

On the other hand, suppose Q is of rank $r < s$. Then Q can be expressed in the form (6.1) with s replaced by r. The n^s copy of this rule is given by

$$\frac{1}{n^s n_1 n_2 \cdots n_r} \sum_{k_s=0}^{n-1} \cdots \sum_{k_1=0}^{n-1} \sum_{j_r=0}^{n_r-1} \cdots \sum_{j_1=0}^{n_1-1}$$

$$\times f\left(\left\{ \frac{(k_1, k_2, \ldots, k_s)}{n} + \frac{j_1}{nn_1}\mathbf{z}_1 + \cdots + \frac{j_r}{nn_r}\mathbf{z}_r \right\}\right).$$

From this expression we see that the quadrature points for this n^s copy include the quadrature points for the n^s-point product-rectangle rule. Since the latter rule has rank s and invariants $n, n, \ldots, n$, it is a simple consequence of Theorem 3.37 that the n^s copy of the original rank-r rule has rank s. (From the part of this theorem already proved, it then follows that the s invariants of the n^s copy are $nn_1, \ldots, nn_r, n, \ldots, n$.) $\blacksquare$

The maximal rank rules that we are concerned with in this chapter are n^s copies of rank-1 lattice rules. If the single invariant of the rank-1 rule is m, then the n^s copy rule has invariants $mn, n, \ldots, n$, with n occurring $s - 1$ times in this list. We shall show in the next section that there exist rules of this type which yield an interesting bound on P_α. As in the rank-1 rule case, this bound is obtained by deriving an expression for a certain mean value of P_α.

In Section 6.4, the expression for this mean is compared with the corresponding mean for rank-1 rules with the same (or approximately the same) number of points. We shall see in the final section that in the case of

2^s copies the ratio of these means decreases exponentially to zero as the dimension $s \to \infty$. This provides a good theoretical basis for the practical use of rules which are 2^s copies of well-chosen rank-1 rules.

6.3 Existence of rules satisfying good bounds on P_α

Starting with the rank-1 rule of order m,

$$Q(\mathbf{z}, m)f = \frac{1}{m} \sum_{j=0}^{m-1} f\left(\left\{\frac{j}{m}\mathbf{z}\right\}\right), \tag{6.3}$$

the n^s copy rule is

$$Q^{(n)}(\mathbf{z}, m)f = \frac{1}{n^s m} \sum_{k_s=0}^{n-1} \cdots \sum_{k_1=0}^{n-1} \sum_{j=0}^{m-1} f\left(\left\{\frac{j}{nm}\mathbf{z} + \frac{(k_1,\ldots,k_s)}{n}\right\}\right). \tag{6.4}$$

We now show that P_α for the copy rule $Q^{(n)}(\mathbf{z}, m)$ can be expressed as an error involving $Q(\mathbf{z}, m)$.

Lemma 6.5. *If $Q^{(n)}(\mathbf{z}, m)$ is the lattice rule given by (6.4), then, for $\alpha > 1$,*

$$P_\alpha\left(Q^{(n)}(\mathbf{z}, m)\right) = Q(\mathbf{z}, m)f_\alpha^{(n)} - 1, \tag{6.5}$$

where

$$f_\alpha^{(n)}(\mathbf{x}) = \prod_{i=1}^{s} F_\alpha^{(n)}(x_i), \tag{6.6}$$

with

$$F_\alpha^{(n)}(x) = 1 + \frac{1}{n^\alpha} \sum_{h \neq 0} \frac{1}{|h|^\alpha} e^{2\pi i h x} = 1 + \frac{1}{n^\alpha}(F_\alpha(x) - 1). \tag{6.7}$$

Proof If $Q(\mathbf{z}, m)$ has the lattice L, then $Q^{(n)}(\mathbf{z}, m)$ has the lattice $n^{-1}L$. But the dual of $n^{-1}L$ is $nL^\perp$. It then follows from the definition of P_α in (5.1) that

$$\begin{aligned}
P_\alpha\left(Q^{(n)}(\mathbf{z}, m)\right) &= \sum_{\mathbf{h} \in nL^\perp} \frac{1}{(\overline{h_1}\overline{h_2}\cdots\overline{h_s})^\alpha} - 1 \\
&= \sum_{\mathbf{h} \in L^\perp} \frac{1}{(\overline{nh_1}\overline{nh_2}\cdots\overline{nh_s})^\alpha} - 1 \\
&= Q(\mathbf{z}, m)f_\alpha^{(n)} - 1,
\end{aligned}$$

where Theorem 2.8 has been used once again, with

$$f_\alpha^{(n)}(\mathbf{x}) := \sum_{\mathbf{h} \in \mathbb{Z}^s} \frac{e^{2\pi i \mathbf{h}\cdot\mathbf{x}}}{(\overline{nh_1}\overline{nh_2}\cdots\overline{nh_s})^\alpha}. \tag{6.8}$$

Clearly, $f_\alpha^{(n)}$ may be written in the form (6.6), with

$$F_\alpha^{(n)}(x) := \sum_{h \in \mathbb{Z}} \frac{e^{2\pi i h x}}{(\,\overline{nh}\,)^\alpha} = 1 + \frac{1}{n^\alpha} \sum_{h \neq 0} \frac{1}{|h|^\alpha} e^{2\pi i h x},$$

which with (4.11) completes the proof. $\blacksquare$

The result in Lemma 6.5 will play an important role in what follows. The lemma is not only useful to us in developing the theory, but also very useful in computer searches, since the P_α error for $Q^{(n)}(\mathbf{z}, m)$ (which is a rule of order $n^s m$) may be computed as an error for $Q(\mathbf{z}, m)$, which has order only m.

We now define a convenient mean of $P_\alpha\left(Q^{(n)}(\mathbf{z}, m)\right)$, analogous to that given in Definition 4.5.

Definition 6.6. *For any integer $m \geqslant 2$, let $Z = Z(m)$ be the set of all $\mathbf{z} \in \mathbb{Z}^s$ whose components z_k, $1 \leqslant k \leqslant s$, are relatively prime to m and satisfy $-m/2 < z_k \leqslant m/2$. The mean of $P_\alpha\left(Q^{(n)}(\mathbf{z}, m)\right)$ over $\mathbf{z} \in Z$ is*

$$M_\alpha^{(n)}(m) := \frac{1}{\varphi(m)^s} \sum_{\mathbf{z} \in Z} P_\alpha\left(Q^{(n)}(\mathbf{z}, m)\right), \quad \alpha > 1. \qquad (6.9)$$

The next result, which is analogous to Theorem 4.8, gives an expression for $M_\alpha^{(n)}(m)$ when m is prime, and from it a simple existence theorem. The theorem is from Disney and Sloan (1992).

Theorem 6.7. *If m is a prime number, then*

$$\begin{aligned}
M_\alpha^{(n)}(m) &= \frac{1}{m}\left(1 + \frac{2\zeta(\alpha)}{n^\alpha}\right)^s \\
&\quad + \frac{m-1}{m}\left(1 - \frac{2(1 - m^{1-\alpha})\zeta(\alpha)}{(m-1)n^\alpha}\right)^s - 1. \qquad (6.10)
\end{aligned}$$

Moreover, there exists a $\mathbf{z} \in Z$ for which $P_\alpha\left(Q^{(n)}(\mathbf{z}, m)\right) \leqslant M_\alpha^{(n)}(m)$.

Proof As in Proposition 4.6, it is obvious that there exists a $\mathbf{z} \in Z$ for which $P_\alpha\left(Q^{(n)}(\mathbf{z}, m)\right)$ is no greater than the mean value $M_\alpha^{(n)}(m)$.

The crucial step in the evaluation of $M_\alpha^{(n)}(m)$ is to use Lemma 6.5 to replace the problem by one involving just a rank-1 rule. After that, the argument is similar to that used in the proof of Theorem 4.8. Thus using (6.3) and (6.5) we write $P_\alpha\left(Q^{(n)}(\mathbf{z}, m)\right)$ as

$$P_\alpha\left(Q^{(n)}(\mathbf{z}, m)\right) = \frac{1}{m} f_\alpha^{(n)}(\mathbf{0}) + \left[\frac{1}{m}\sum_{j=1}^{m-1} f_\alpha^{(n)}\left(\left\{\frac{j}{m}\mathbf{z}\right\}\right)\right] - 1,$$

where the $j = 0$ term in the quadrature sum has again been separated out because it is independent of $\mathbf{z}$. Now from (6.6) and (6.7) we find

$$f_\alpha^{(n)}(\mathbf{0}) = \prod_{i=1}^{s} \left(1 + \frac{1}{n^\alpha} \sum_{h \neq 0} \frac{1}{|h|^\alpha} \right) = \left(1 + \frac{2\zeta(\alpha)}{n^\alpha} \right)^s,$$

and therefore

$$P_\alpha\left(Q^{(n)}(\mathbf{z}, m) \right) = \frac{1}{m} \left(1 + \frac{2\zeta(\alpha)}{n^\alpha} \right)^s + \left[\frac{1}{m} \sum_{j=1}^{m-1} f_\alpha^{(n)}\left(\left\{ \frac{j}{m}\mathbf{z} \right\} \right) \right] - 1.$$

Since m is prime, $\varphi(m) = m - 1$. Thus from the definition of $M_\alpha^{(n)}(m)$ in (6.9), we have

$$
\begin{aligned}
M_\alpha^{(n)}(m) \;=\;& \frac{1}{(m-1)^s} \sum_{\mathbf{z} \in Z} \left[\frac{1}{m} \left(1 + \frac{2\zeta(\alpha)}{n^\alpha} \right)^s \right. \\
&\left. + \left[\frac{1}{m} \sum_{j=1}^{m-1} f_\alpha^{(n)}\left(\left\{ \frac{j}{m}\mathbf{z} \right\} \right) \right] - 1 \right] \\
\;=\;& \frac{1}{m} \left(1 + \frac{2\zeta(\alpha)}{n^\alpha} \right)^s + \left[\frac{1}{m(m-1)^s} \sum_{\mathbf{z} \in Z} \sum_{j=1}^{m-1} f_\alpha^{(n)}\left(\left\{ \frac{j}{m}\mathbf{z} \right\} \right) \right] \\
& - 1,
\end{aligned}
$$

in which the means of the first and third terms, which are independent of $\mathbf{z}$, are obtained trivially. Writing this expression as

$$M_\alpha^{(n)}(m) = \frac{1}{m} \left(1 + \frac{2\zeta(\alpha)}{n^\alpha} \right)^s + \theta_\alpha^{(n)}(m) - 1,$$

we then obtain from (6.8)

$$
\begin{aligned}
\theta_\alpha^{(n)}(m) \;=\;& \frac{1}{m(m-1)^s} \sum_{j=1}^{m-1} \sum_{\mathbf{z} \in Z} \sum_{\mathbf{h} \in \mathbb{Z}^s} \frac{e^{2\pi i j \mathbf{h} \cdot \mathbf{z}/m}}{(\overline{nh_1}\,\overline{nh_2} \cdots \overline{nh_s})^\alpha} \\
\;=\;& \frac{1}{m(m-1)^s} \sum_{j=1}^{m-1} \sideset{}{'}\sum_{-m/2 < z_1 \leqslant m/2} \cdots \sideset{}{'}\sum_{-m/2 < z_s \leqslant m/2} \\
& \sum_{h_1 = -\infty}^{\infty} \cdots \sum_{h_s = -\infty}^{\infty} \frac{e^{2\pi i j h_1 z_1/m} \cdots e^{2\pi i j h_s z_s/m}}{(\overline{nh_1}\,\overline{nh_2} \cdots \overline{nh_s})^\alpha}
\end{aligned}
$$

$$= \frac{1}{m} \sum_{j=1}^{m-1} \left(\frac{1}{m-1} {\sum_{-m/2<z\leqslant m/2}}' \sum_{h=-\infty}^{\infty} \frac{e^{2\pi i jhz/m}}{(\overline{nh})^{\alpha}} \right)^{s}.$$

Thus, we can write

$$\theta_{\alpha}^{(n)}(m) = \frac{1}{m} \sum_{j=1}^{m-1} \left(\frac{1}{m-1} T_{\alpha}^{(n)}(j,m) \right)^{s}, \qquad (6.11)$$

where

$$T_{\alpha}^{(n)}(j,m) = \sum_{h=-\infty}^{\infty} {\sum_{-m/2<z\leqslant m/2}}' \frac{e^{2\pi i jhz/m}}{(\overline{nh})^{\alpha}}, \qquad 1 \leqslant j \leqslant m-1.$$

Now, separating out the terms with h a multiple of m, we obtain

$$\frac{1}{m-1} T_{\alpha}^{(n)}(j,m) = \frac{1}{m-1} \left(\sum_{k=-\infty}^{\infty} {\sum_{-m/2<z\leqslant m/2}}' \frac{1}{(\overline{mnk})^{\alpha}} \right.$$

$$\left. + \sum_{h\not\equiv 0\,(\mathrm{mod}\,m)} \frac{1}{|nh|^{\alpha}} {\sum_{-m/2<z\leqslant m/2}}' e^{2\pi i jhz/m} \right)$$

$$= 1 + \frac{2\zeta(\alpha)}{n^{\alpha}m^{\alpha}}$$

$$+ \frac{1}{m-1} \sum_{h\not\equiv 0\,(\mathrm{mod}\,m)} \frac{1}{|nh|^{\alpha}} {\sum_{-m/2<z\leqslant m/2}}' e^{2\pi i jhz/m}.$$

But for $h \not\equiv 0\,(\mathrm{mod}\,m)$ and $1 \leqslant j \leqslant m-1$ we have

$${\sum_{-m/2<z\leqslant m/2}}' e^{2\pi i jhz/m} = \sum_{-m/2<z\leqslant m/2} e^{2\pi i jhz/m} - 1 = -1,$$

since jh is not a multiple of m for j in this range and m prime. Thus

$$\frac{1}{m-1} T_{\alpha}^{(n)}(j,m) = 1 + \frac{2\zeta(\alpha)}{n^{\alpha}m^{\alpha}} + \frac{1}{m-1} \sum_{h\not\equiv 0\,(\mathrm{mod}\,m)} \frac{-1}{|nh|^{\alpha}}$$

$$= 1 + \frac{2\zeta(\alpha)}{n^{\alpha}m^{\alpha}} - \frac{1}{(m-1)n^{\alpha}} \left(\sum_{h\neq 0} \frac{1}{|h|^{\alpha}} - \sum_{k\neq 0} \frac{1}{|mk|^{\alpha}} \right)$$

$$= 1 + \frac{2\zeta(\alpha)}{n^{\alpha}m^{\alpha}} - \frac{2}{(m-1)n^{\alpha}} \left(\zeta(\alpha) - \frac{\zeta(\alpha)}{m^{\alpha}} \right)$$

$$= 1 - \frac{2(1 - m^{1-\alpha})\zeta(\alpha)}{(m-1)n^\alpha}.$$

Since $T_\alpha^{(n)}(j, m)$ turns out to be independent of j, substitution of this last expression into (6.11) yields

$$\theta_\alpha^{(n)}(m) = \frac{m-1}{m}\left(1 - \frac{2(1 - m^{1-\alpha})\zeta(\alpha)}{(m-1)n^\alpha}\right)^s,$$

and hence the desired result. ∎

6.4   Comparison with rank-1 rules

In this section we compare the quantity $M_\alpha^{(n)}(m)$ with the corresponding mean for rank-1 rules. Now Theorem 4.8 shows that, for N prime, the mean of P_α for rank-1 rules of order N is given by

$$M_\alpha(N) = \frac{(1 + 2\zeta(\alpha))^s}{N} + \frac{N-1}{N}\left(1 - \frac{2(1 - N^{1-\alpha})\zeta(\alpha)}{N-1}\right)^s - 1. \quad (6.12)$$

Since we want to compare $M_\alpha^{(n)}(m)$ with the mean for rank-1 rules of comparable order, let $\widehat{M}_\alpha(n^s m)$ be the value obtained by taking $N = n^s m$ in the right-hand side of (6.12). This is not the true value of $M_\alpha(n^s m)$ because (6.12) is only valid when N is prime, but $\widehat{M}_\alpha(n^s m)$ can be regarded as an approximation to the value which $M_\alpha(N)$ would take if N were prime and close to $n^s m$. The values of $\widehat{M}_\alpha(n^s m)$ and $M_\alpha^{(n)}(m)$ will be compared for prime values of m. To do this, let

$$\Omega = \Omega_\alpha^{(n)}(m) = \frac{M_\alpha^{(n)}(m)}{\widehat{M}_\alpha(n^s m)}, \quad (6.13)$$

which is the quotient

$$n^s\left(1 + \frac{2\zeta(\alpha)}{n^\alpha}\right)^s + n^s(m-1)\left(1 - \frac{2(1 - m^{1-\alpha})\zeta(\alpha)}{(m-1)n^\alpha}\right)^s - n^s m \quad (6.14)$$

divided by

$$(1 + 2\zeta(\alpha))^s + (n^s m - 1)\left(1 - \frac{2(1 - (n^s m)^{1-\alpha})\zeta(\alpha)}{n^s m - 1}\right)^s - n^s m. \quad (6.15)$$

Also, let

$$\lambda = \frac{n + 2\zeta(\alpha)/n^{\alpha-1}}{1 + 2\zeta(\alpha)}, \quad (6.16)$$

so that λ^s is the ratio of the first terms in the numerator and denominator.

We want Ω to be small. Theorem 6.9 below shows that, under suitable conditions, Ω is bounded by λ^s. Given that fact, the following proposition taken from Disney and Sloan (1992) motivates the choice of $n = 2$.

Proposition 6.8. *Let λ be given by (6.16).*

(i) *If $n = 2$, then $\lambda < 1$ for all $\alpha > 1$.*
(ii) *For any fixed $\alpha > 1$, the minimum value of λ occurs when $n = 2$.*

Proof To prove part (i), we first note that when $n = 2$ the inequality $\lambda < 1$ is equivalent to $2\zeta(\alpha) > (1 - 2^{1-\alpha})^{-1}$, which holds for all $\alpha \geqslant 2$ since $\zeta(\alpha) > 1$ for all $\alpha > 1$ and $(1 - 2^{1-\alpha})^{-1} \leqslant 2$ when $\alpha \geqslant 2$. Because $\zeta(\alpha) > (\alpha-1)^{-1}$ for all $\alpha > 1$, the proof can be completed by showing that $2(\alpha - 1)^{-1} \geqslant (1 - 2^{1-\alpha})^{-1}$ for $1 < \alpha < 2$. This is equivalent to $f(x) \geqslant 0$ for $0 < x < 1$, where $f(x) = 1 - 2^{-x} - x/2$. Now $f(0) = f(1) = 0$ while $f''(x) = -2^{-x}(\log 2)^2 < 0$. This implies that $f(x) \geqslant 0$ for $0 \leqslant x \leqslant 1$. Hence $\lambda < 1$ for $\alpha > 1$.

To prove part (ii), for $x \geqslant 1$ let

$$f(x) = \frac{x + 2x^{1-\alpha}\zeta(\alpha)}{1 + 2\zeta(\alpha)}.$$

It has to be shown that the minimum of f over positive integer values of x occurs at $x = 2$. Since $f(1) = 1$ and part (i) shows that $f(2) < 1$, it is sufficient to show that $f'(x) > 0$ for $x \geqslant 3$ and that $f(3) > f(2)$.

By differentiating f, it may be seen that $f'(x)$ is positive if and only if $\zeta(\alpha) < x^\alpha/(2(\alpha - 1))$. But it follows from (4.40) that $\zeta(\alpha) < \alpha/(\alpha - 1)$ for $\alpha > 1$. Thus $f'(x)$ is positive if $x^\alpha - 2\alpha > 0$. This inequality holds at $\alpha = 1$ for all $x > 2$ and the derivative of the left-hand side with respect to α is $x^\alpha \log x - 2$, which is positive for all $\alpha > 1$, $x \geqslant 3$. Thus $f'(x) > 0$ for $x \geqslant 3$.

To complete the proof, we need to show that $f(3) > f(2)$ for all $\alpha > 1$. This is equivalent to showing that $3^{\alpha-1} - 2(c^{\alpha-1} - 1)\zeta(\alpha) > 0$, where $c = 3/2$. Since $\zeta(\alpha) < \alpha/(\alpha - 1)$, it need only be shown that $g(x) > 0$ for $x > 0$, where $g(x) = x3^x - 2(x + 1)(c^x - 1)$. But $g(0) = 0$, $g'(0) > 0$ and $g''(x)$ is positive for all $x > 0$, since the sign of $g''(x)$ is the same as the sign of

$$2^x - 2\left(\frac{(x + 1)\log^2 c + 2\log c}{x\log^2 3 + 2\log 3}\right).$$

This expression is positive at $x = 0$, and if $a = \log 3$ and $b = \log c$ then its derivative is

$$2^x \log 2 + \frac{2b(2(a - b) + ab)}{a(ax + 2)^2},$$

which is positive. $\blacksquare$

The next result shows that Ω is small when λ^s is small. It is based on Disney and Sloan (1992, Theorem 5), but is more general in that it is not restricted to $n = 2$, and less general in that it does not cover the $1 < \alpha < 2$ case. Readers interested in this case should consult the original reference.

Theorem 6.9. *Let* $\Omega = \Omega_\alpha^{(n)}(m)$ *and* λ *be given by* (6.13) *and* (6.16), *respectively, and suppose* m *is a prime number. Assume also that either* $s \geqslant 3$ *and* $\alpha \geqslant 2$, *or* $s = 2$ *and* $\alpha \geqslant 3$. *If* $n = 2$ *then* $\Omega < \lambda^s$. *If* $n \geqslant 3$ *then* $\Omega < \lambda^s$ *for all* m *sufficiently large.*

Proof Let

$$
\begin{aligned}
t_1 &= \left(n + \frac{2\zeta(\alpha)}{n^{\alpha-1}}\right)^s, \\
t_2 &= n^s(m-1)\left(1 - \frac{2(1 - m^{1-\alpha})\zeta(\alpha)}{(m-1)n^\alpha}\right)^s, \\
b_1 &= (1 + 2\zeta(\alpha))^s, \\
b_2 &= (n^s m - 1)\left(1 - \frac{2(1 - (n^s m)^{1-\alpha})\zeta(\alpha)}{n^s m - 1}\right)^s,
\end{aligned}
$$

and $\qquad c = n^s m.$

Then proving the result $\Omega < \lambda^s$ is equivalent to proving that

$$
\frac{t_1 + t_2 - c}{b_1 + b_2 - c} < \frac{t_1}{b_1}.
$$

Since $b_1 > 0$ and $c > 0$, this is equivalent to

$$
b_1(1 - t_2/c) - t_1(1 - b_2/c) > 0, \tag{6.17}
$$

provided $b_1 + b_2 - c > 0$.

Now, if

$$
\hat{b}_2 = (n^s m - 1)\left(1 - \frac{2\zeta(\alpha)}{n^s m - 1}\right)^s,
$$

then we have

$$
\begin{aligned}
b_1 + \hat{b}_2 - c &= (1 + 2\zeta(\alpha))^s + (n^s m - 1)\left(1 - \frac{2\zeta(\alpha)}{n^s m - 1}\right)^s - n^s m \\
&= \sum_{j=0}^{s} \binom{s}{j}(2\zeta(\alpha))^j \\
&\quad + (n^s m - 1)\sum_{j=0}^{s}\binom{s}{j}(-1)^j \left(\frac{2\zeta(\alpha)}{n^s m - 1}\right)^j - n^s m \\
&= \sum_{j=0}^{s} \binom{s}{j}(2\zeta(\alpha))^j \left(1 + (-1)^j (n^s m - 1)^{1-j}\right) - n^s m
\end{aligned}
$$

$$= \sum_{j=1}^{s} \binom{s}{j} (2\zeta(\alpha))^j \left(1 + (-1)^j (n^s m - 1)^{1-j}\right) > 0.$$

Thus $b_1 + b_2 - c > 0$ if $b_2 > \hat{b}_2$. This last inequality clearly holds for all $n \geqslant 2$ if $1 - 2\zeta(\alpha)/(2^s m - 1)$ is positive, the latter condition being equivalent to $1 + 2\zeta(\alpha) < 2^s m$. For $\alpha \geqslant 2$ this holds for all $m \geqslant 2$ and $s \geqslant 2$, since $1 + 2\zeta(\alpha) \leqslant 1 + 2\pi^2/6 < 5 < 2^s m$. Note that under the same conditions $\hat{b}_2 > 0$.

So all that remains is to prove (6.17). Since $\alpha \geqslant 2$, $s \geqslant 2$, and $m \geqslant 2$, we have

$$0 < \frac{2(1 - m^{1-\alpha})\zeta(\alpha)}{n^\alpha (m - 1)} \leqslant \frac{\zeta(2)}{2} < 1.$$

This implies that $t_2 < n^s(m-1)$, and hence $1 - t_2/c > 1/m$. Since we also have $b_2 > \hat{b}_2$, it then follows that (6.17) holds if

$$b_1 - t_1 m(1 - \hat{b}_2/c) > 0. \tag{6.18}$$

Now fix $\alpha \geqslant 2$, $s \geqslant 2$, and $n \geqslant 2$, and consider the left-hand side of (6.18) as a function $g(m)$ for $m \geqslant 1$. If $n = 2$, then $g(1) \geqslant b_1 - t_1 > 0$, since $\hat{b}_2 > 0$ and $b_1 - t_1 > 0$ by Proposition 6.8(i). Also, for all $n \geqslant 2$ $g(m) \to b_1 - n^{-s}t_1(1 + 2s\zeta(\alpha))$ as $m \to \infty$. At any stationary point of g,

$$\frac{-g'(m)}{t_1} = 1 - \left(1 - \frac{2\zeta(\alpha)}{n^s m - 1}\right)^s - \left(1 - \frac{2\zeta(\alpha)}{n^s m - 1}\right)^{s-1} \frac{2s\zeta(\alpha)}{n^s m - 1} = 0. \tag{6.19}$$

Substitution for $(1 - 2\zeta(\alpha)/(n^s m - 1))^s$ from (6.19) yields

$$\begin{aligned} g(m) &= b_1 - n^{-s}t_1(1 + 2s\zeta(\alpha)(1 - 2\zeta(\alpha)/(n^s m - 1))^{s-1}) \\ &> b_1 - n^{-s}t_1(1 + 2s\zeta(\alpha)) = g(\infty) \end{aligned}$$

at any stationary point of g.

It follows that if $g(\infty)$ is positive, then $g(m)$ is positive for all m if $n = 2$, and for all sufficiently large m if $n \geqslant 3$. Thus it is sufficient to establish $g(\infty) > 0$, or, equivalently,

$$\left(\frac{1 + 2\zeta(\alpha)}{1 + 2n^{-\alpha}\zeta(\alpha)}\right)^s > 1 + 2s\zeta(\alpha).$$

Obviously, this result holds for all $n \geqslant 2$ if it holds for $n = 2$. Restricting attention to this case, the condition to be established can be written as

$$\left(\frac{1 + 2\zeta(\alpha)}{2 + 2^{2-\alpha}\zeta(\alpha)}\right)^s - \frac{1 + 2s\zeta(\alpha)}{2^s} > 0. \tag{6.20}$$

Let $u(\alpha, s)$ denote the left-hand side of (6.20). Then u is an increasing function of s since the first term of (6.20) is an increasing function of s by

Proposition 6.8(i), and the second term is increasing for $s \geqslant 2$. Thus (6.20) holds for $s \geqslant 3$ if it holds for $s = 3$.

Let $s = 3$ and suppose $\alpha \geqslant 2$. Then

$$
\begin{aligned}
\left(2 + 2^{2-\alpha}\zeta(\alpha)\right)^3 u(\alpha,3) &= (1 + 2\zeta(\alpha))^3 - (1 + 6\zeta(\alpha))(1 + 2^{1-\alpha}\zeta(\alpha))^3 \\
&\geqslant (1 + 2\zeta(\alpha))^3 - (1 + 6\zeta(\alpha))(1 + \zeta(\alpha)/2)^3 \\
&= -3\zeta(2\zeta^3 - 9\zeta^2 - 6\zeta + 4)/8,
\end{aligned}
$$

where $\zeta = \zeta(\alpha)$. Elementary reasoning shows that this polynomial in ζ is positive at least for $1 \leqslant \zeta \leqslant 2$. Since $1 < \zeta(\alpha) \leqslant \pi^2/6 < 2$ for $\alpha \geqslant 2$, then $u(\alpha, 3) > 0$ for $\alpha \geqslant 2$. This proves the result for $s \geqslant 3$, $\alpha \geqslant 2$. Similarly,

$$
\left(2 + 2^{2-\alpha}\zeta(\alpha)\right)^2 u(\alpha, 2) \geqslant -\zeta(4\zeta^2 - 31\zeta + 8)/16
$$

for $\alpha \geqslant 3$. This polynomial is positive for $1 \leqslant \zeta \leqslant 2$, which proves the result for $s = 2$, $\alpha \geqslant 3$. $\blacksquare$

6.5 What is special about 2^s copies?

After the rigours of the preceding section, we present here a less forbidding summary of the main results, and attempt an informal explanation.

The aim, let us recall, is to compare n^s copies of rank-1 rules with uncopied rank-1 rules having a similar number of points. To this end we make use of the ratio

$$
\Omega_\alpha^{(n)}(m) = \frac{M_\alpha^{(n)}(m)}{\widehat{M}_\alpha(n^s m)}. \tag{6.21}
$$

Here the numerator $M_\alpha^{(n)}(m)$ is an appropriate mean of our standard error measure P_α, applied to n^s copies of the m-point rank-1 rule

$$
\frac{1}{m}\sum_{j=0}^{m-1} f\left(\left\{\frac{j}{m}\mathbf{z}\right\}\right),
$$

with m taken to be prime. (The mean is taken over the set of $\mathbf{z}$ vectors defined in Definition 6.6.) The denominator is the corresponding mean for rank-1 rules with a similar (but prime) total number of points $N \approx n^s m$. If n^s copies are to have a useful role then this ratio should be less than 1, or at least close to 1.

Both the numerator and the denominator of (6.21) are available explicitly, but the expressions are complicated. However, Theorem 6.9 shows that, for $\alpha \geqslant 2$, $s \geqslant 3$, and m sufficiently large,

$$
\Omega_\alpha^{(n)}(m) < \left(\frac{n + 2\zeta(\alpha)/n^{\alpha-1}}{1 + 2\zeta(\alpha)}\right)^s. \tag{6.22}
$$

For example, setting $\alpha = 2$ we obtain

$$\Omega_2^{(n)}(m) < \left(\frac{n + \pi^2/(3n)}{1 + \pi^2/3} \right)^s. \tag{6.23}$$

With numerical values inserted, the bound becomes

$$\Omega_2^{(2)}(m) < (0.85)^s$$

for $n = 2$ (in which case the bound holds for all prime values of m),

$$\Omega_2^{(3)}(m) < (0.95)^s$$

for $n = 3$, and

$$\Omega_2^{(4)}(m) < (1.12)^s$$

for $n = 4$. (These bounds conform, of course, with Proposition 6.8, which asserts that the right-hand side of (6.23) is < 1 for $n = 2$, and takes its minimum value for $n = 2$.)

How can we understand the bound (6.22)? It is clear that it is obtained by retaining just the dominant first terms in (6.14) and (6.15), or, equivalently, by using the approximations

$$M_\alpha^{(n)}(m) \approx \frac{1}{m} \left(1 + \frac{2\zeta(\alpha)}{n^\alpha} \right)^s, \tag{6.24}$$

$$\widehat{M}_\alpha(n^s m) \approx \frac{1}{n^s m} \left(1 + 2\zeta(\alpha) \right)^s. \tag{6.25}$$

Where do these dominant first terms come from? Recall that $\widehat{M}_\alpha(N)$ (with N a prime number near $n^s m$) is the mean over a set of vectors $\mathbf{z}$ of

$$
\begin{aligned}
P_\alpha(\mathbf{z}, N) &= Q(\mathbf{z}, N) f_\alpha - 1 \\
&= \frac{1}{N} \sum_{j=0}^{N-1} f_\alpha \left(\left\{ \frac{j}{N} \mathbf{z} \right\} \right) - 1 \\
&= \frac{1}{N} f_\alpha(\mathbf{0}) + \frac{1}{N} \sum_{j=1}^{N-1} f_\alpha \left(\left\{ \frac{j}{N} \mathbf{z} \right\} \right) - 1.
\end{aligned}
$$

It is the first term of this expression (which is of course independent of $\mathbf{z}$) that gives the approximation (6.25), since $f_\alpha(\mathbf{0}) = (1 + 2\zeta(\alpha))^s$ (see (4.10) and (4.11)). Thus the surprising conclusion is that, when s is large, *the principal source of the mean error for rank-1 rules with $N \approx n^s m$ points is the large contribution to the quadrature sum from the large value of our standard test function f_α at the vertices of the unit cube.*

Similarly, on making use of Lemma 6.5, $M_\alpha^{(n)}(m)$ is the mean of

$$P_\alpha\left(Q^{(n)}(\mathbf{z}, m)\right) = \frac{1}{m} f_\alpha^{(n)}(\mathbf{0}) + \frac{1}{m}\sum_{j=1}^{m-1} f_\alpha^{(n)}\left(\left\{\frac{j}{m}\mathbf{z}\right\}\right) - 1,$$

the first term of which gives (6.24).

Thus we arrive at the following interpretation of the approximations (6.24) and (6.25), and hence of the bound (6.22). One effect of taking n^s copies of rank-1 rules is to increase the mean error by a factor of n^s (for in effect Lemma 6.5 has us using just m points instead of $n^s m$ points), but this is offset, especially when n is small, by a significant reduction in $f_\alpha^{(n)}(\mathbf{0}) = (1 + 2\zeta(\alpha)/n^\alpha)^s$, the vertex value of the effective test integrand $f_\alpha^{(n)}$, as n increases. In a delicate balancing act, the best result turns out to be achieved with $n = 2$. (Of course we should not be surprised that it is not a good idea to take the value of n to be large: the result of copying many times is a rule that is more and more like the product-rectangle rule, which is a poor rule when s is large.)

A final word of caution: the results we have been discussing depend critically on the definition of P_α. It would take only modest changes in that definition to change the balance in favour of not copying at all (that can happen if we make the test function f_α easier, by reducing the vertex peaks), or of copying more than twice (if we make f_α even more peaked at the vertices).

Perhaps a balanced conclusion is that the results in this chapter at least give hope that 2^s copies of rank-1 rules can compete, as far as accuracy is concerned, with uncopied rank-1 rules. This question is explored in numerical experiments in the last chapter of this book. By then we will have seen that 2^s copies also have an advantage of a quite different kind: namely, that they can yield not only an estimate of the integral, but also, at no extra cost, an estimate of the error.

7

Intermediate rank lattice rules

7.1 Introduction

The previous chapter considered lattice rules of maximal rank. Now in this chapter we shall look at certain lattice rules that have rank less than or equal to s. In particular, we are concerned with rules of the form

$$Q_r f = \frac{1}{n^r m} \sum_{k_r=0}^{n-1} \cdots \sum_{k_1=0}^{n-1} \sum_{j=0}^{m-1} f\left(\left\{\frac{j}{m}\mathbf{z} + \frac{k_1}{n}\mathbf{y}_1 + \cdots + \frac{k_r}{n}\mathbf{y}_r\right\}\right), \quad (7.1)$$

for $0 \leqslant r \leqslant s$, where n and m are relatively prime and $\mathbf{z}, \mathbf{y}_1, \ldots, \mathbf{y}_r \in \mathbb{Z}^s$. Such lattice rules will be used in Chapter 10 to produce embedded sequences of lattice rules as r increases from 0 to s.

Under certain conditions on the integer vectors $\mathbf{z}, \mathbf{y}_1, \ldots, \mathbf{y}_r$, the points in (7.1) are distinct, so that Q_r is a lattice rule of order $n^r m$. Moreover, unless $r = 0$, it is a rule of rank r. This result is stated below. In the rest of this chapter we shall always assume that the conditions given in this theorem hold.

Theorem 7.1. *Let n and m be relatively prime, and suppose that $\mathbf{z} \in \mathbb{Z}^s$ has no nontrivial factor in common with m. Also suppose that the set*

$$\left\{\left\{\frac{k_1}{n}\mathbf{y}_1 + \cdots + \frac{k_r}{n}\mathbf{y}_r\right\} : 0 \leqslant k_p < n, \quad 1 \leqslant p \leqslant r\right\} \quad (7.2)$$

contains n^r distinct points. Then the lattice rule Q_r, $0 \leqslant r \leqslant s$, given by (7.1) has order $n^r m$. Moreover, Q_0 is a rank-1 rule with single invariant m, while for $1 \leqslant r \leqslant s$, Q_r is a rank-r lattice rule with invariants $nm, \underbrace{n, \ldots, n}_{r-1 \text{ times}}$.

Proof Because m and n^r are relatively prime, the rule Q_r is the direct sum (see Section 3.3) of Q_0 and q_r, where

$$Q_0 f = \frac{1}{m} \sum_{j=0}^{m-1} f\left(\left\{\frac{j}{m}\mathbf{z}\right\}\right), \quad (7.3)$$

and

$$q_r f = \frac{1}{n^r} \sum_{k_r=0}^{n-1} \cdots \sum_{k_1=0}^{n-1} f\left(\left\{\frac{k_1}{n}\mathbf{y}_1 + \cdots + \frac{k_r}{n}\mathbf{y}_r\right\}\right).$$

The assumptions in the theorem ensure that Q_0 is a rank-1 rule of order m and q_r is a rank-r rule with invariants $n, n, \ldots, n$. A direct application of Theorem 3.29 then tells us that for $r \geqslant 1$ Q_r is a rank-r rule with invariants $mn, n, \ldots, n$. ∎

For the case $r = 2$, Q_2 is simply a rank-2 lattice rule expressed in the three-sum form of (3.29). This form has already been discussed extensively in Section 3.6 and used in Section 5.4.

We see from Theorem 7.1 that Q_s is a lattice rule having maximal rank. Hence we know from Theorem 6.4 that Q_s is the n^s copy of a rule of lower rank. It turns out that Q_s is the n^s copy of Q_0 (Joe and Sloan 1992a). A proof of this result for a particular choice of $\mathbf{y}_1, \ldots, \mathbf{y}_s$ may be found in Section 10.2 (see Proposition 10.2).

For a particular choice of the integer vectors $\mathbf{y}_1, \ldots, \mathbf{y}_r$, it is possible to obtain bounds on P_α for lattice rules of the form (7.1). These bounds are derived in the next section. The results of this section are then used in the final section of this chapter to compare these lattice rules with rank-1 rules.

7.2 Existence of good rank-r rules

Bounds on P_α for well-chosen rank-1 rules and maximal rank rules were obtained in Sections 4.5 and 6.3 respectively. These bounds were derived by finding an expression for a certain mean value of P_α. In this section we intend to show that well-chosen rank-r rules Q_r lead to a similar bound on $P_\alpha(Q_r)$ for $1 \leqslant r \leqslant s - 1$.

A new difficulty here is that Q_r is dependent on $r + 1$ integer vectors, namely, $\mathbf{z}, \mathbf{y}_1, \ldots, \mathbf{y}_r$, whereas the rank-1 and maximal rank lattice rules looked at earlier depended on only one integer vector. If we wanted to obtain a mean value of $P_\alpha(Q_r)$ by taking a mean over all admissible $\mathbf{z}, \mathbf{y}_1, \ldots, \mathbf{y}_r$, then the problem would be complicated. To overcome this difficulty, we fix the integer vectors $\mathbf{y}_1, \ldots, \mathbf{y}_r$ in a special way, and then obtain an expression for the mean of $P_\alpha(Q_r)$ over all admissible $\mathbf{z}$. Analogously to Definition 4.5, the mean is then defined by:

Definition 7.2. *Let the integer vectors $\mathbf{y}_1, \ldots, \mathbf{y}_r$ be fixed. Also, for any integer $m \geqslant 2$, let $Z = Z(m)$ be the set of all $\mathbf{z} \in \mathbb{Z}^s$ whose components z_k are relatively prime to m and satisfy $-m/2 < z_k \leqslant m/2$. The mean of $P_\alpha(Q_r)$ over $\mathbf{z} \in Z$ is*

$$M_{\alpha,r}^{(n)}(m) := \frac{1}{\varphi(m)^s} \sum_{\mathbf{z} \in Z} P_\alpha(Q_r), \quad \alpha > 1. \tag{7.4}$$

Though the conditions in Theorem 7.1 place restrictions on the integer vectors $\mathbf{y}_1, \ldots, \mathbf{y}_r$, there are obviously still many possible choices for these vectors. However, as discussed in Section 5.4, the numerical results of Sloan and Walsh (1990) indicated that in the case where $r = 2$ with $n = 2$, good choices for $\mathbf{y}_1$ and $\mathbf{y}_2$ were $\mathbf{y}_1 = (1, 0, 0, \ldots, 0)$ and $\mathbf{y}_2 = (0, 1, 0, \ldots, 0)$. This encourages us to make an analogous choice for all the vectors $\mathbf{y}_1, \ldots, \mathbf{y}_r$. Thus we choose $\mathbf{y}_p$, $1 \leqslant p \leqslant r$, to be the vector with all components set equal to 0 except for the pth component, which is set equal to 1. With this choice, we then have

$$Q_r f = \frac{1}{n^r m} \sum_{k_r=0}^{n-1} \cdots \sum_{k_1=0}^{n-1} \sum_{j=0}^{m-1} f\left(\left\{ \frac{j}{m}\mathbf{z} + \frac{(k_1, \ldots, k_r, 0, \ldots, 0)}{n} \right\}\right). \tag{7.5}$$

It turns out that this choice allows us to calculate $P_\alpha(Q_r)$, for any fixed r, with only m (rather than $n^r m$) function evaluations, because $P_\alpha(Q_r)$ is the error in a certain m-point rule applied to a modified function. With hindsight, some of the computer searches of rank-2 lattice rules made by Sloan and Walsh (1990) could have been done in a quarter of the time if this result had been known at that time. The result, which generalizes Lemma 6.5, is not only of practical use, but will also be useful to us in obtaining an expression for the mean of $P_\alpha(Q_r)$. It first appeared in Joe and Sloan (1992a).

Theorem 7.3. *For $\alpha > 1$, $1 \leqslant r \leqslant s$, and $n \geqslant 2$, let $f_{\alpha,r}^{(n)}$ be the function on $\mathbb{R}^s$ defined by*

$$f_{\alpha,r}^{(n)}(\mathbf{x}) := \left(\prod_{p=1}^{r} F_\alpha^{(n)}(x_p) \right) \prod_{q=r+1}^{s} F_\alpha(x_q), \tag{7.6}$$

where, as in (6.7) and (4.11),

$$F_\alpha^{(n)}(x) = 1 + \frac{1}{n^\alpha} \sum_{h \neq 0} \frac{1}{|h|^\alpha} e^{2\pi \imath h x} \quad \text{and} \quad F_\alpha(x) = 1 + \sum_{h \neq 0} \frac{1}{|h|^\alpha} e^{2\pi \imath h x}. \tag{7.7}$$

If $Q_r^{(n)} f$ is the m-point lattice rule defined by

$$Q_r^{(n)} f := \frac{1}{m} \sum_{j=0}^{m-1} f\left(\left\{ \left(\frac{jn}{m} z_1, \ldots, \frac{jn}{m} z_r, \frac{j}{m} z_{r+1}, \ldots, \frac{j}{m} z_s \right) \right\}\right), \tag{7.8}$$

then

$$P_\alpha(Q_r) = Q_r^{(n)} f_{\alpha,r}^{(n)} - 1. \tag{7.9}$$

Proof From (5.2) and (7.5) we have

$$
\begin{aligned}
P_\alpha(Q_r) \;=\;& \frac{1}{n^r m} \sum_{k_r=0}^{n-1} \cdots \sum_{k_1=0}^{n-1} \sum_{j=0}^{m-1} \\
& \times f_\alpha\left(\left\{\frac{j}{m}\mathbf{z} + \frac{(k_1,\ldots,k_r,0,\ldots,0)}{n}\right\}\right) - 1 \\
=\;& \frac{1}{n^r m} \sum_{k_r=0}^{n-1} \cdots \sum_{k_1=0}^{n-1} \sum_{j=0}^{m-1} \left[\left(\prod_{p=1}^{r} \sum_{h=-\infty}^{\infty} \frac{1}{\overline{h}^\alpha} e^{2\pi i h(j z_p/m + k_p/n)}\right)\right. \\
& \left.\times \prod_{q=r+1}^{s} F_\alpha\left(\left\{\frac{j}{m} z_q\right\}\right)\right] - 1 \\
=\;& \frac{1}{m} \sum_{j=0}^{m-1} K_{j,r}^{(n)} \prod_{q=r+1}^{s} F_\alpha\left(\left\{\frac{j}{m} z_q\right\}\right) - 1,
\end{aligned}
$$

where we have used (4.10) in the second step, and in the third

$$
\begin{aligned}
K_{j,r}^{(n)} \;:=\;& \frac{1}{n^r} \sum_{k_r=0}^{n-1} \cdots \sum_{k_1=0}^{n-1} \prod_{p=1}^{r} \left(1 + \sum_{h\neq 0} \frac{1}{|h|^\alpha} e^{2\pi i h(j z_p/m + k_p/n)}\right) \\
=\;& \prod_{p=1}^{r} \frac{1}{n} \sum_{k=0}^{n-1} \left(1 + \sum_{h\neq 0} \frac{1}{|h|^\alpha} e^{2\pi i h j z_p/m} (e^{2\pi i h/n})^k\right) \\
=\;& \prod_{p=1}^{r} \frac{1}{n} \left(n + \sum_{\substack{h\neq 0 \\ h\equiv 0\,(\mathrm{mod}\,n)}} \frac{1}{|h|^\alpha} e^{2\pi i h j z_p/m}\, n\right) \\
=\;& \prod_{p=1}^{r} \left(1 + \frac{1}{n^\alpha} \sum_{k\neq 0} \frac{1}{|k|^\alpha} e^{2\pi i k j n z_p/m}\right) \\
=\;& \prod_{p=1}^{r} F_\alpha^{(n)}\left(\left\{\frac{jn}{m} z_p\right\}\right).
\end{aligned}
$$

Thus

$$
P_\alpha(Q_r) = \frac{1}{m} \sum_{j=0}^{m-1} \left(\prod_{p=1}^{r} F_\alpha^{(n)}\left(\left\{\frac{jn}{m} z_p\right\}\right)\right) \prod_{q=r+1}^{s} F_\alpha\left(\left\{\frac{j}{m} z_q\right\}\right) - 1,
$$

which is equivalent to (7.9). ∎

In the case when $r = s$ we see that Theorem 7.3 yields

$$
P_\alpha(Q_s) = \frac{1}{m} \sum_{j=0}^{m-1} f_{\alpha,s}^{(n)}\left(\left\{\frac{jn}{m}\mathbf{z}\right\}\right) - 1.
$$

This is equivalent to (6.5) obtained earlier, since the cyclic group generated by $\mathbf{z}/m$ is the same, when n and m are relatively prime, as the cyclic group generated by $n\mathbf{z}/m$.

For later use we note that, from (7.6) and (7.7),

$$f_{\alpha,r}^{(n)}(\mathbf{x}) = \sum_{\mathbf{h}\in\mathbb{Z}^s} \frac{1}{\left(\prod\limits_{p=1}^{r} \overline{nh_p} \prod\limits_{q=r+1}^{s} \overline{h_q}\right)^{\alpha}} \, e^{2\pi\imath \mathbf{h}\cdot\mathbf{x}}. \tag{7.10}$$

We now obtain an expression for $M_{\alpha,r}^{(n)}(m)$ for the case when m is prime, and deduce from it that there exists a $\mathbf{z}$ for which $P_\alpha(Q_r)$ has a good bound. This result was proved by Joe and Disney (1993). We remark that it would be possible, but not easy, to also obtain an expression for $M_{\alpha,r}^{(n)}(m)$ when m is not prime (compare Theorem 4.10).

Theorem 7.4. *For $\alpha > 1$, $1 \leqslant r \leqslant s$, m a prime number, and n not a multiple of m,*

$$\begin{aligned}
M_{\alpha,r}^{(n)}(m) \;=\;& \frac{1}{m}\left(1 + \frac{2\zeta(\alpha)}{n^\alpha}\right)^r (1 + 2\zeta(\alpha))^{s-r} \\[2mm]
&+ \frac{m-1}{m}\left(1 - \frac{2(1 - m^{1-\alpha})\zeta(\alpha)}{(m-1)n^\alpha}\right)^r \\[2mm]
&\times \left(1 - \frac{2(1 - m^{1-\alpha})\zeta(\alpha)}{m-1}\right)^{s-r} - 1.
\end{aligned}$$

Moreover, there exists a $\mathbf{z} \in Z$ for which $P_\alpha(Q_r) \leqslant M_{\alpha,r}^{(n)}(m)$.

Proof It is clear that there exists a $\mathbf{z} \in Z$ for which $P_\alpha(Q_r)$ is less than or equal to $M_{\alpha,r}^{(n)}(m)$, the mean of $P_\alpha(Q_r)$ over Z (compare Proposition 4.6). To complete the proof, we only need to prove the expression for $M_{\alpha,r}^{(n)}(m)$ given above. Given $\mathbf{z} = (z_1, \ldots, z_s)$, let $\mathbf{w}$ be the vector

$$\mathbf{w} = (nz_1, \ldots, nz_r, z_{r+1}, \ldots, z_s). \tag{7.11}$$

Using (7.8) and (7.9), we can then write $P_\alpha(Q_r)$ as

$$P_\alpha(Q_r) = \frac{1}{m}f_{\alpha,r}^{(n)}(\mathbf{0}) + \left[\frac{1}{m}\sum_{j=1}^{m-1} f_{\alpha,r}^{(n)}\left(\left\{\frac{j}{m}\mathbf{w}\right\}\right)\right] - 1,$$

where the $j = 0$ term in the quadrature sum has been separated out since it is independent of $\mathbf{w}$. Now from (7.6) and (7.7) we have

$$
\begin{aligned}
f_{\alpha,r}^{(n)}(\mathbf{0}) &= \left[\prod_{p=1}^{r}\left(1+\frac{1}{n^\alpha}\sum_{h\neq 0}\frac{1}{|h|^\alpha}\right)\right]\prod_{q=r+1}^{s}\left(1+\sum_{h\neq 0}\frac{1}{|h|^\alpha}\right)\\
&= \left(1+\frac{2\zeta(\alpha)}{n^\alpha}\right)^{r}(1+2\zeta(\alpha))^{s-r},
\end{aligned}
\tag{7.12}
$$

and hence

$$
P_\alpha(Q_r) = \frac{1}{m}\left(1+\frac{2\zeta(\alpha)}{n^\alpha}\right)^{r}(1+2\zeta(\alpha))^{s-r}+\left[\frac{1}{m}\sum_{j=1}^{m-1}f_{\alpha,r}^{(n)}\left(\left\{\frac{j}{m}\mathbf{w}\right\}\right)\right]-1.
$$

Since m is prime, we have $\varphi(m) = m - 1$. If W_r is the set of all $\mathbf{w}$ of the form (7.11) with $\mathbf{z} \in Z$, then it follows from the definition of $M_{\alpha,r}^{(n)}(m)$ in (7.4) that

$$
\begin{aligned}
M_{\alpha,r}^{(n)}(m) &= \frac{1}{(m-1)^s}\sum_{\mathbf{w}\in W_r}\left[\frac{1}{m}\left(1+\frac{2\zeta(\alpha)}{n^\alpha}\right)^{r}(1+2\zeta(\alpha))^{s-r}\right.\\
&\qquad\left.+\left[\frac{1}{m}\sum_{j=1}^{m-1}f_{\alpha,r}^{(n)}\left(\left\{\frac{j}{m}\mathbf{w}\right\}\right)\right]-1\right]\\
&= \frac{1}{m}\left(1+\frac{2\zeta(\alpha)}{n^\alpha}\right)^{r}(1+2\zeta(\alpha))^{s-r}\\
&\qquad+\left[\frac{1}{m(m-1)^s}\sum_{\mathbf{w}\in W_r}\sum_{j=1}^{m-1}f_{\alpha,r}^{(n)}\left(\left\{\frac{j}{m}\mathbf{w}\right\}\right)\right]-1,
\end{aligned}
$$

in which the means of the first and third terms are obtained trivially since they are independent of $\mathbf{w}$. Writing this as

$$
M_{\alpha,r}^{(n)}(m) = \frac{1}{m}\left(1+\frac{2\zeta(\alpha)}{n^\alpha}\right)^{r}(1+2\zeta(\alpha))^{s-r}+\theta_{\alpha,r}^{(n)}(m)-1,
$$

and defining $\Lambda = \Lambda(m) := \{z \in \mathbb{Z} : -m/2 < z \leqslant m/2,\ z \neq 0\}$, we then obtain from (7.10)

$$
\begin{aligned}
\theta_{\alpha,r}^{(n)}(m) &= \frac{1}{m(m-1)^s}\sum_{j=1}^{m-1}\sum_{\mathbf{w}\in W_r}\sum_{\mathbf{h}\in\mathbb{Z}^s}\frac{e^{2\pi i j\mathbf{h}\cdot\mathbf{w}/m}}{\left(\prod_{p=1}^{r}\overline{nh_p}\prod_{q=r+1}^{s}\overline{h_q}\right)^\alpha}\\
&= \frac{1}{m(m-1)^s}\sum_{j=1}^{m-1}\sum_{z_1\in\Lambda}\cdots\sum_{z_s\in\Lambda}\\
&\qquad\sum_{h_1=-\infty}^{\infty}\cdots\sum_{h_s=-\infty}^{\infty}\prod_{p=1}^{r}\frac{e^{2\pi i jnh_p z_p/m}}{(\overline{nh_p})^\alpha}\prod_{q=r+1}^{s}\frac{e^{2\pi i jh_q z_q/m}}{\overline{h_q}^\alpha}
\end{aligned}
$$

$$= \frac{1}{m} \sum_{j=1}^{m-1} \left(\frac{1}{m-1} \sum_{h=-\infty}^{\infty} \sum_{z\in\Lambda} \frac{e^{2\pi ijnhz/m}}{(\overline{nh})^{\alpha}} \right)^{r}$$

$$\times \left(\frac{1}{m-1} \sum_{h=-\infty}^{\infty} \sum_{z\in\Lambda} \frac{e^{2\pi ijhz/m}}{\overline{h}^{\alpha}} \right)^{s-r}.$$

Thus we can write

$$\theta_{\alpha,r}^{(n)}(m) = \frac{1}{m} \sum_{j=1}^{m-1} \left(\frac{1}{m-1} T_{\alpha,r}^{(n)}(j,m) \right)^{r} \left(\frac{1}{m-1} T_{\alpha,r}^{(1)}(j,m) \right)^{s-r}, \quad (7.13)$$

where for any positive integer k

$$T_{\alpha,r}^{(k)}(j,m) = \sum_{h=-\infty}^{\infty} \sum_{z\in\Lambda} \frac{e^{2\pi ijkhz/m}}{(\overline{kh})^{\alpha}}, \quad 1 \leqslant j \leqslant m-1.$$

Separating out the terms with $h \equiv 0 \,(\mathrm{mod}\, m)$, we obtain

$$T_{\alpha,r}^{(n)}(j,m) = \sum_{h=-\infty}^{\infty} \sum_{z\in\Lambda} \frac{1}{(\overline{mnh})^{\alpha}} + \sum_{h\not\equiv 0\,(\mathrm{mod}\,m)} \frac{1}{|nh|^{\alpha}} \sum_{z\in\Lambda} e^{2\pi ijnhz/m}$$

$$= (m-1)\left(1 + \frac{2\zeta(\alpha)}{m^{\alpha}n^{\alpha}}\right) + \sum_{h\not\equiv 0\,(\mathrm{mod}\,m)} \frac{1}{|nh|^{\alpha}} \sum_{z\in\Lambda} e^{2\pi ijnhz/m}.$$

But for $h \not\equiv 0 \,(\mathrm{mod}\, m)$, we have

$$\sum_{z\in\Lambda} e^{2\pi ijnhz/m} = \sum_{-m/2 < z \leqslant m/2} e^{2\pi ijnhz/m} - 1 = -1,$$

since m is prime and n is not a multiple of m, so that jhn is not a multiple of m for $1 \leqslant j \leqslant m-1$. Thus we obtain

$$\frac{1}{m-1} T_{\alpha,r}^{(n)}(j,m) = 1 + \frac{2\zeta(\alpha)}{m^{\alpha}n^{\alpha}} + \frac{1}{m-1} \sum_{h\not\equiv 0\,(\mathrm{mod}\,m)} \frac{-1}{|nh|^{\alpha}}$$

$$= 1 + \frac{2\zeta(\alpha)}{m^{\alpha}n^{\alpha}} - \frac{1}{(m-1)n^{\alpha}} \left(\sum_{h\neq 0} \frac{1}{|h|^{\alpha}} - \sum_{k\neq 0} \frac{1}{|mk|^{\alpha}} \right)$$

$$= 1 + \frac{2\zeta(\alpha)}{m^{\alpha}n^{\alpha}} - \frac{2}{(m-1)n^{\alpha}} \left(\zeta(\alpha) - \frac{1}{m^{\alpha}}\zeta(\alpha) \right)$$

$$= 1 - \frac{2(1 - m^{1-\alpha})\zeta(\alpha)}{(m-1)n^{\alpha}}. \quad (7.14)$$

Setting $n = 1$,

$$\frac{1}{m-1} T_{\alpha,r}^{(1)}(j,m) = 1 - \frac{2(1 - m^{1-\alpha})\zeta(\alpha)}{m-1}. \quad (7.15)$$

Since $T_{\alpha,r}^{(n)}(j,m)$ and $T_{\alpha,r}^{(1)}(j,m)$, $1 \leqslant j \leqslant m-1$, are independent of j, we see from (7.13) that (7.14) and (7.15) yield the desired result. $\blacksquare$

We remark that the result also holds for $n=1$ or $r=0$. In either case, the right-hand side of (7.5) is just a rank-1 rule and $M_{\alpha,r}^{(n)}(m)$ becomes the mean of P_α for rank-1 rules of prime order m. This recovers the result found in Theorem 4.8 (with $N=m$). If $n \geqslant 2$ and $r=s$, then, as expected, we obtain the result given in Theorem 6.7.

7.3 Comparison with rank-1 rules

We now compare the quantity $M_{\alpha,r}^{(n)}(m)$ with the corresponding mean for rank-1 rules. Theorem 4.8 shows that, for N prime, the mean of P_α for rank-1 rules of order N is given by

$$M_\alpha(N) = \frac{(1 + 2\zeta(\alpha))^s}{N} + \frac{N-1}{N}\left(1 - \frac{2(1 - N^{1-\alpha})\zeta(\alpha)}{N-1}\right)^s - 1. \quad (7.16)$$

Since we want to compare $M_{\alpha,r}^{(n)}(m)$ with the mean for rank-1 rules of comparable order, as in Section 6.4, let $\widehat{M_\alpha}(n^r m)$ be the value obtained by taking $N = n^r m$ in the right-hand side of (7.16). This is not the true value of $M_\alpha(n^r m)$ because (7.16) is only valid when N is prime, but $\widehat{M_\alpha}(n^r m)$ can be regarded as an approximation to the value which $M_\alpha(N)$ would take if N were prime and close to $n^r m$. Thus we look at the ratio

$$\Omega = \Omega_{\alpha,r}^{(n)}(m) = \frac{M_{\alpha,r}^{(n)}(m)}{\widehat{M_\alpha}(n^r m)}, \quad (7.17)$$

which is the quotient of

$$n^r\left(1 + \frac{2\zeta(\alpha)}{n^\alpha}\right)^r (1 + 2\zeta(\alpha))^{s-r}$$
$$+ n^r(m-1)\left(1 - \frac{2(1 - m^{1-\alpha})\zeta(\alpha)}{(m-1)n^\alpha}\right)^r \left(1 - \frac{2(1 - m^{1-\alpha})\zeta(\alpha)}{m-1}\right)^{s-r}$$
$$- n^r m$$

and

$$(1 + 2\zeta(\alpha))^s + (n^r m - 1)\left(1 - \frac{2(1 - (n^r m)^{1-\alpha})\zeta(\alpha)}{n^r m - 1}\right)^s - n^r m.$$

Clearly, we want to show that $\Omega < 1$. As we shall see below, Ω is small when the quantity λ given in (6.16) is small. The properties of λ were studied earlier in the previous chapter. In particular Proposition 6.8 shows that for any fixed $\alpha > 1$, the minimum value of λ occurs when $n=2$. This motivates the choice of $n=2$. We now show that $\Omega < \lambda^r$ when $n=2$.

Theorem 7.5. *Let Ω and λ be given by (7.17) and (6.16), respectively, and suppose m is a prime number. Assume $n = 2$, $s \geqslant 2$, $\alpha \geqslant 2$, $m \geqslant 5$, and $1 \leqslant r \leqslant s - 1$. Then $\Omega < \lambda^r$.*

Proof If b_1, b_2, t_1, t_2, and c are positive real numbers, then it is not difficult to verify that

$$\frac{t_1 + t_2 - c}{b_1 + b_2 - c} < \frac{t_1}{b_1}$$

if

$$b_1 > t_1, \qquad b_1 + b_2 > c, \qquad \text{and } t_2 < b_2 < c. \tag{7.18}$$

Thus the result is proved if we can prove that (7.18) holds for $n = 2$ when

$$
\begin{aligned}
t_1 &= n^r \left(1 + \frac{2\zeta(\alpha)}{n^\alpha}\right)^r (1 + 2\zeta(\alpha))^{s-r}, \\
t_2 &= n^r(m-1) \\
&\quad \times \left(1 - \frac{2(1 - m^{1-\alpha})\zeta(\alpha)}{(m-1)n^\alpha}\right)^r \left(1 - \frac{2(1 - m^{1-\alpha})\zeta(\alpha)}{m-1}\right)^{s-r}, \\
b_1 &= (1 + 2\zeta(\alpha))^s, \\
b_2 &= (n^r m - 1)\left(1 - \frac{2(1 - (n^r m)^{1-\alpha})\zeta(\alpha)}{n^r m - 1}\right)^s,
\end{aligned}
$$

and $\quad c = n^r m$.

It may not be clear at first glance that t_2 and b_2 are positive quantities, but it can be seen that a sufficient condition for this to hold is that $m \geqslant 2\zeta(\alpha) + 1$. This is satisfied since $m \geqslant 5 > \pi^2/3 + 1$ and $\pi^2/3 \geqslant 2\zeta(\alpha)$ for $\alpha \geqslant 2$. For $n = 2$, we see from Proposition 6.8 that $t_1/b_1 = \lambda^r < 1$, so $b_1 > t_1$.

Now we note that $b_2 \geqslant \hat{b}_2$, where

$$\hat{b}_2 = (n^r m - 1)\left(1 - \frac{2\zeta(\alpha)}{n^r m - 1}\right)^s,$$

and we have

$$
\begin{aligned}
b_1 + b_2 - c &\geqslant b_1 + \hat{b}_2 - c \\
&= (1 + 2\zeta(\alpha))^s + (n^r m - 1)\left(1 - \frac{2\zeta(\alpha)}{n^r m - 1}\right)^s - n^r m \\
&= \sum_{j=0}^{s} \binom{s}{j} (2\zeta(\alpha))^j \\
&\quad + (n^r m - 1)\sum_{j=0}^{s} \binom{s}{j}(-1)^j \left(\frac{2\zeta(\alpha)}{n^r m - 1}\right)^j - n^r m
\end{aligned}
$$

$$= \sum_{j=0}^{s} \binom{s}{j} (2\zeta(\alpha))^j \left(1 + (-1)^j (n^r m - 1)^{1-j}\right) - n^r m$$

$$= \sum_{j=1}^{s} \binom{s}{j} (2\zeta(\alpha))^j \left(1 + (-1)^j (n^r m - 1)^{1-j}\right) > 0.$$

It is clear that $b_2 < c$ and so all that remains is to prove that $t_2 < b_2$. Now for $1 \leqslant r \leqslant s - 1$, if

$$\hat{t}_2 = n^r (m - 1) \left(1 - \frac{2(1 - m^{1-\alpha})\zeta(\alpha)}{m - 1}\right)^{s-r},$$

then we see that $t_2 \leqslant \hat{t}_2$. Moreover we have

$$n^r (m - 1) \left(1 - \frac{2\zeta(\alpha)}{n^r m - 1}\right)^s < (n^r m - 1) \left(1 - \frac{2\zeta(\alpha)}{n^r m - 1}\right)^s = \hat{b}_2 \leqslant b_2.$$

Thus a sufficient condition for $t_2 < b_2$ to hold is if

$$\hat{t}_2 \leqslant n^r (m - 1) \left(1 - \frac{2\zeta(\alpha)}{n^r m - 1}\right)^s,$$

that is, if

$$\left(1 - \frac{2(1 - m^{1-\alpha})\zeta(\alpha)}{m - 1}\right)^{s-r} \leqslant \left(1 - \frac{2\zeta(\alpha)}{n^r m - 1}\right)^s. \tag{7.19}$$

We remark that (7.19) does not hold when $r = s$, so the proof given here does not extend to the case where $r = s$. Recalling that $m > 2\zeta(\alpha) + 1$ for $\alpha \geqslant 2$, it follows that $0 < 2(1 - m^{1-\alpha})\zeta(\alpha)/(m - 1) < 1$, and that $0 < 2\zeta(\alpha)/(n^r m - 1) < 1$. Since

$$\log(1 - x) = -\sum_{h=1}^{\infty} \frac{x^h}{h} \quad \text{for } |x| < 1,$$

on taking the natural logarithms of both sides of (7.19), we see that an equivalent condition is

$$-(s - r) \sum_{h=1}^{\infty} \frac{\left(\dfrac{2(1 - m^{1-\alpha})\zeta(\alpha)}{m - 1}\right)^h}{h} \leqslant -s \sum_{h=1}^{\infty} \frac{\left(\dfrac{2\zeta(\alpha)}{n^r m - 1}\right)^h}{h}.$$

So the proof is complete once we prove for $h \geqslant 1$ that

$$\frac{s^{1/h}}{n^r m - 1} \leqslant \frac{(s - r)^{1/h}(1 - m^{1-\alpha})}{m - 1}. \tag{7.20}$$

Now it is easily seen that $(m-1)/(n^r m - 1) \leqslant n^{-r}$ and thus (7.20) holds if, for $h \geqslant 1$,

$$n^{-r} \leqslant \left(\frac{s-r}{s}\right)^{1/h}(1 - m^{1-\alpha}),$$

which is equivalent to the inequality

$$m \geqslant \frac{1}{\left(1 - n^{-r}\left(\frac{s}{s-r}\right)^{1/h}\right)^{\frac{1}{\alpha-1}}}. \tag{7.21}$$

A straightforward consideration of the derivative of $g(r) := n^{-r}s/(s-r)$ shows that g is a nonincreasing function of r for $1 \leqslant r \leqslant s-2$, while $g(s-2) \geqslant g(s-1)$ (with equality for $n=2$), and so for $h \geqslant 1$, we have

$$
\begin{aligned}
n^{-r}\left(\frac{s}{s-r}\right)^{1/h} &\leqslant n^{-r}\frac{s}{s-r} \\
&\leqslant n^{-1}\frac{s}{s-1} \\
&\leqslant (3/2)n^{-1} \text{ when } s \geqslant 3 \\
&= 3/4
\end{aligned}
$$

for $n=2$. Thus for $s \geqslant 3$ and $n=2$ (7.21) holds when $m \geqslant 4^{1/(\alpha-1)}$. Since $4^{1/(\alpha-1)} \leqslant 4$ when $\alpha \geqslant 2$, and $m \geqslant 5$ by assumption, this completes the proof when $s \geqslant 3$.

For the case where $s=2$ (and so $r=1$), we need to work with (7.19) directly. With $n=2$, this yields the inequality

$$\frac{-2(1-m^{1-\alpha})\zeta(\alpha)}{m-1} \leqslant \frac{-4\zeta(\alpha)}{2m-1} + \frac{(2\zeta(\alpha))^2}{(2m-1)^2}, \tag{7.22}$$

which is equivalent to

$$\zeta(\alpha) \geqslant (2m-1) - \frac{(2m-1)^2(1-m^{1-\alpha})}{2(m-1)}.$$

Since $\alpha \geqslant 2$, then

$$
\begin{aligned}
(2m-1) - \frac{(2m-1)^2(1-m^{1-\alpha})}{2(m-1)} &\leqslant (2m-1) - \frac{(2m-1)^2(1-m^{-1})}{2(m-1)} \\
&= \frac{2m-1}{2m} < 1.
\end{aligned}
$$

As $\zeta(\alpha) > 1$ for all $\alpha \geqslant 2$, we see that (7.22) is satisfied, completing the proof for $s=2$. $\blacksquare$

In the last theorem, we have assumed that $\alpha \geqslant 2$. This is not usually a serious restriction, as in practice $\alpha > 1$ is normally taken to be an even integer. This, as indicated earlier, allows us to use the Bernoulli polynomials of degree α to calculate P_α. However, it is possible to obtain a similar result for $1 < \alpha < 2$, though this result is not as general as that in the preceding theorem. Interested readers may find such a result in Joe and Disney (1993).

In the particular case where $r = 2$ and $n = 2$ the rule (7.5) becomes

$$
Q_2 f = \frac{1}{4m} \sum_{k_2=0}^{1} \sum_{k_1=0}^{1} \sum_{j=0}^{m-1} f\left(\left\{\frac{j}{m}\mathbf{z} + \frac{(k_1, k_2, 0, \ldots, 0)}{n}\right\}\right). \tag{7.23}
$$

Theorem 7.5 tells us that for m prime the mean of P_α for $\mathbf{z} \in Z(m)$, divided by the corresponding mean for rank-1 rules, is bounded by

$$
\Omega_{\alpha,2}^{(2)}(m) \leqslant \left(\frac{2 + \zeta(\alpha)/2^{\alpha-2}}{1 + 2\zeta(\alpha)}\right)^2,
$$

provided $s \geqslant 2$, $\alpha \geqslant 2$, and $m \geqslant 5$. For example, if $\alpha = 2$ the bound becomes

$$
\Omega_{2,2}^{(2)}(m) \leqslant \left(\frac{2 + \pi^2/6}{1 + \pi^2/3}\right)^2 = 0.72.
$$

When m is large the expressions on the right of these bounds are good approximations to Ω.

The experimental results of Sloan and Walsh (1990), discussed in Section 5.4, provide an interesting sidelight on these bounds: we saw that the ratio of the *best* value of P_2 for rank-2 rules with invariants $2m$, 2 to that of the *best* value of P_2 for rank-1 rules with a similar number of points was always less than 1, and typically was around 0.9 (see, for example, Figure 5.2). Moreover, the best rank-2 values were almost invariably obtained using rules of the form (7.23). The experimental observations for $r = 2$ therefore give some confidence that theoretical results for the ratios of mean values of P_2 can be indicators of the likely ratios of best values of P_2 from the two classes; in other words, that the probability distributions in the two cases seem more or less the same. Further numerical evidence supporting this may be found in Joe and Disney (1993).

8
Lattice rules for nonperiodic integrands

8.1 Introduction

So far we have assumed that the integral to be evaluated is

$$If = \int_{C^s} f(\mathbf{x})d\mathbf{x},$$

where f has a continuous (and preferably smooth) one-periodic extension with respect to each component of $\mathbf{x}$. Because such a periodic property does not often arise naturally, in Section 2.12 we showed how to force periodization. But forced periodization has its problems: there are significant overhead costs, and what is perhaps worse, forced periodization by the nonlinear transformation method has the nasty feature that it turns easy integrals into harder ones. (Even the constant integrand $f \equiv 1$ presents a challenge to the lattice rule after the periodization process!)

In this chapter we consider a different alternative: if the given integrand f is continuous on C^s but does not have a continuous one-periodic extension, the lattice rule (with a few modifications) is now applied directly, without any nonlinear transformation. The presentation follows Niederreiter and Sloan (1993, in press). The final section gives some advice on when the direct approach might be appropriate. (The reader should be warned that the application of lattice rules to nonperiodic integrands is in its infancy, and that any conclusions must be treated with caution!)

We suppose in this chapter that

$$Qf = \frac{1}{N} \sum_{j=0}^{N-1} f(\mathbf{x}_j) \tag{8.1}$$

is a lattice rule, with the particular property that, apart from $\mathbf{x}_0 = \mathbf{0}$, every other quadrature point lies in $(0,1)^s$, the interior of the unit cube. The following proposition characterizes the lattice rules with this property.

Proposition 8.1. *For an N-point lattice rule Q, every quadrature point other than $\mathbf{0}$ lies in the interior of the unit cube if and only if Q is a rank-1 rule in which the generating integer vector $\mathbf{z}$ in (4.1) has each component relatively prime to N.*

Proof Suppose that Q is the rank-1 rule (4.1), with each component of $\mathbf{z}$ relatively prime to N. Then the one-dimensional projections of the quadrature points $\{\mathbf{z}/N\}, \{2\mathbf{z}/N\}, \ldots, \{(N-1)\mathbf{z}/N\}$ on any axis are just $1/N, 2/N, \ldots, (N-1)/N$ in some order. On the other hand, if some component of $\mathbf{z}$ (say z_k) is not relatively prime to N, then the corresponding one-dimensional projection has fewer than N distinct points. Since (as explained in Section 3.4) each point in that projected rule must occur the same number of times, there is therefore at least one quadrature point $\mathbf{x} \neq \mathbf{0}$ that projects to zero, that is to say, there is at least one quadrature point other than $\mathbf{0}$ lying on the boundary plane $x_k = 0$. If Q is a rule with rank $r \geqslant 2$ written in the canonical form (3.3), then because n_{i+1} divides n_i for $1 \leqslant i \leqslant r-1$ it is clear that each one-dimensional projection of Q has n_1 or fewer points, where n_1 is the largest of the invariants. Since

$$n_1 = \frac{N}{n_2 \cdots n_r} < N \text{ if } r > 1,$$

it follows, by the argument above, that there is a quadrature point other than $\mathbf{0}$ on each boundary plane $x_k = 0$, $1 \leqslant k \leqslant s$. $\blacksquare$

It is certainly possible to use formula (8.1) just as it stands, but now that the periodicity assumption is abandoned, we might first want to think about this question: *why are we favouring the vertex of the cube at the origin over all other vertices?*

To focus our thoughts, consider the one-dimensional case. The lattice rule Q then becomes the rectangle rule

$$Qf = \frac{1}{N} \sum_{j=0}^{N-1} f\left(\frac{j}{N}\right).$$

As long as f is periodic, this rule has the same effect as the trapezoidal rule

$$\frac{1}{2N}[f(0) + f(1)] + \frac{1}{N} \sum_{j=1}^{N-1} f\left(\frac{j}{N}\right),$$

so there is no need to make a distinction. But once we allow nonperiodic functions the rectangle rule and trapezoidal rule have quite different effects. The universally preferred choice out of these two is the trapezoidal rule, because, unlike the rectangle rule, it integrates all linear functions exactly.

With this example in mind, Niederreiter and Sloan (1993) introduced a class of 'vertex-modified' lattice rules corresponding to the rule (8.1). These are rules of the form

$$Mf = \sum_{i_s=0}^{1} \cdots \sum_{i_1=0}^{1} w_{i_1 \cdots i_s} f(i_1, \ldots, i_s) + \frac{1}{N} \sum_{j=1}^{N-1} f(\mathbf{x}_j), \qquad (8.2)$$

so that $w_{i_1\cdots i_s}$, with $i_k = 0$ or 1 for $k = 1,\ldots,s$, is the quadrature weight corresponding to the vertex $(i_1,\ldots,i_s)$ of the unit cube. The weights are assumed to satisfy

$$\sum_{i_s=0}^{1} \cdots \sum_{i_1=0}^{1} w_{i_1\cdots i_s} = \frac{1}{N}, \tag{8.3}$$

ensuring that at least the constant functions are integrated exactly. We see immediately that $Mf = Qf$ if $f(\mathbf{x})$ is one-periodic with respect to each component of $\mathbf{x}$. (Use the fact that f then has the same value at every vertex of the unit cube.)

For example, if $s = 2$ the vertex-modified rules are rules of the form

$$Mf = w_{00}f(0,0) + w_{01}f(0,1) + w_{10}f(1,0) + w_{11}f(1,1) + \frac{1}{N}\sum_{j=1}^{N} f(\mathbf{x}_j),$$

with

$$w_{00} + w_{01} + w_{10} + w_{11} = \frac{1}{N}.$$

The two-dimensional case is discussed in more detail in Section 8.6.

We shall see that there is more than one sensible way to choose the vertex weights in the rule (8.2) while still satisfying (8.3). To each such choice corresponds a value of the 'vertex variance'

$$v(M) := \sum_{i_s=0}^{1} \cdots \sum_{i_1=0}^{1} w_{i_1\cdots i_s}^2. \tag{8.4}$$

The lattice rule Q that we started with is obtained by setting $w_{0\cdots 0} = 1/N$ and taking all other vertex weights equal to 0; thus the vertex variance in this case is

$$v(Q) = \frac{1}{N^2}.$$

A large value of the vertex variance signals a potential problem with numerical stability. This is clearly not a significant issue in the case of rule Q, since in this rule the vertex weight at the origin is no greater than that at any interior point, but soon we shall encounter other rules for which the vertex variance is enormous. It will turn out that the large vertex variance of those rules is associated with troubles other than numerical stability, and that the vertex variance is therefore a valuable diagnostic tool.

8.2 Two analogues of the trapezoidal rule

Niederreiter and Sloan (1993) point out that two quite different vertex-modified rules could claim to be generalizations of the one-dimensional trapezoidal rule. The first is suggested by the form of the trapezoidal rule, the second by its property of integrating linear functions exactly.

The 'symmetric' vertex-modified lattice rule follows the one-dimensional example in taking the weights at all vertices to be equal. Remembering that there are 2^s vertices and that the sum of the vertex weights is $1/N$, the definition is

$$Tf := \frac{1}{2^s N} \sum_{i_s=0}^{1} \cdots \sum_{i_1=0}^{1} f(i_1, \ldots, i_s) + \frac{1}{N} \sum_{j=1}^{N-1} f(\mathbf{x}_j). \tag{8.5}$$

This choice gives the least possible value of the vertex variance, namely,

$$v(T) = \frac{1}{2^s N^2}. \tag{8.6}$$

The symmetric rule also turns out to be interesting theoretically, in that with this choice the error expression in Theorem 2.8, which gives the lattice rule error as a sum of Fourier coefficients of f, still holds in a certain sense. The main difference is that we can no longer assume that the Fourier series of f is absolutely convergent (since if this were the case then f would have a continuous periodic extension). For this reason we must now be careful to say exactly what we mean by an infinite sum of Fourier coefficients. The following result is proved (with somewhat greater generality) by Price and Sloan (1992). For notational simplicity the result is stated here just for $s = 2$, but analogous results hold equally in all dimensions.

Theorem 8.2. *Let f be continuous on C^2, and assume that the partial derivatives $\frac{\partial f}{\partial x}$, $\frac{\partial f}{\partial y}$, and $\frac{\partial^2 f}{\partial x \partial y}$ are continuous on C^2, where the partial derivatives at the boundary of the unit square are to be understood in an appropriate one-sided sense. Then*

$$Tf - If = \lim_{n \to \infty} \sideset{}{'}\sum_{\substack{\mathbf{h} \in L^\perp \\ |h_1|, |h_2| \leqslant n}} \hat{f}(\mathbf{h}). \tag{8.7}$$

Like the one-dimensional trapezoidal rule, the rule T is exact for all linear functions:

Theorem 8.3. *If the rule (8.1) is a lattice rule in which every quadrature point other than $\mathbf{0}$ lies in $(0,1)^s$, then the symmetric vertex-modified rule T given by (8.5) is exact for all functions of the form*

$$f(x_1, \ldots, x_s) = a + \sum_{i=1}^{s} b_i x_i.$$

Proof Since rule T is obviously exact for a constant, it is sufficient to prove the result for each linear monomial x_i, $i = 1, \ldots, s$. Since x_i involves only a single component of $\mathbf{x}$, the effect of T when applied to x_i is the same as that of the one-dimensional projection of T on to the x_i axis (see Definition 3.33). But that projection is just the one-dimensional $(N+1)$-point trapezoidal rule. (Remember that the projection of any lattice rule is a lattice rule, and that we have assumed that the only quadrature points with one-dimensional projections projecting on to 0 are the vertices of the cube.) We already know that the one-dimensional trapezoidal rule is exact for all linear functions, so the proof is complete.   ∎

If the symmetric rule T has all those desirable properties, why would we consider anything else? Consider the case $s = 2$. Rule T integrates the function $a + bx + cy$ exactly, but (unlike, for example, the product-trapezoidal rule) it does not in general integrate the bilinear function xy exactly. In s dimensions the comparison with the product-trapezoidal rule becomes even more unfavourable, in that the latter integrates every multi-linear function, including, for example, $x_1 x_2 \cdots x_s$.

With this in mind Niederreiter and Sloan (1993) proposed a second vertex-modified rule, which we shall denote by B, whose defining property is that it integrates every possible multilinear function exactly: *rule B is defined by*

$$Bf := \sum_{i_s=0}^{1} \cdots \sum_{i_1=0}^{1} W_{i_1\cdots i_s} f(i_1,\ldots,i_s) + \frac{1}{N} \sum_{j=1}^{N-1} f(\mathbf{x}_j), \qquad (8.8)$$

where the weights $W_{i_1\cdots i_s}$ are such that B integrates exactly every monomial of the form $x_{j_1} x_{j_2} \cdots x_{j_t}$ with $1 \leqslant j_1 < j_2 < \cdots < j_t \leqslant s$ and $t \leqslant s$.

Niederreiter and Sloan (1993) call this rule the 'optimal' rule, for a reason which will become clear in Section 8.5.

In practice the weights $W_{i_1\cdots i_s}$ are easily found from the formula (8.10) below.

8.3   Vertex weights for the 'optimal' rule B

For simplicity we consider first the case where $s = 2$. Recall that the defining property of rule B is that it integrates every multilinear (or in this case bilinear) function exactly. An obvious basis for the space of bilinear functions is $\{1, x, y, xy\}$, but it turns out that a more convenient basis is $\{\ell_{00}, \ell_{10}, \ell_{01}, \ell_{11}\}$, where

$$\ell_{00}(x,y) = (1-x)(1-y), \quad \ell_{10}(x,y) = x(1-y),$$

$$\ell_{01}(x,y) = (1-x)y, \quad \ell_{11}(x,y) = xy.$$

This basis has the property that ℓ_{ik}, for $i, k = 0$ or 1, has the value 1 at the vertex (i, k) and the value 0 at every other vertex; for example, the function ℓ_{11} is illustrated in Figure 8.1.

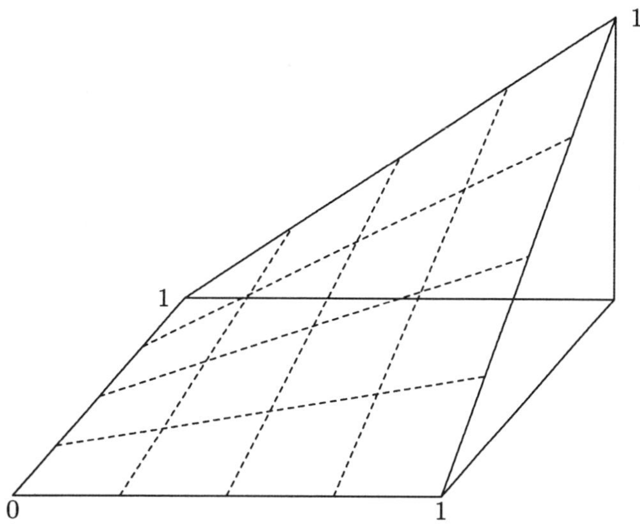

Fig. 8.1 ℓ_{11}.

On applying rule B to the bilinear function ℓ_{ik}, we find

$$
\begin{aligned}
B\ell_{ik} &= \sum_{m=0}^{1}\sum_{j=0}^{1} W_{jm}\ell_{ik}(j,m) + \frac{1}{N}\sum_{j=1}^{N-1}\ell_{ik}(\mathbf{x}_j) \\
&= W_{ik} + \frac{1}{N}\sum_{j=1}^{N-1}\ell_{ik}(\mathbf{x}_j),
\end{aligned}
$$

which can be solved for W_{ik}. Since B gives the exact result applied to every bilinear function we have $B\ell_{ik} = I\ell_{ik} = 2^{-2}$, and hence

$$
W_{ik} = \frac{1}{2^2} - \frac{1}{N}\sum_{j=1}^{N-1}\ell_{ik}(\mathbf{x}_j), \qquad i,k = 0 \text{ or } 1. \tag{8.9}
$$

A similar argument holds in any number of dimensions s. Thus we define

$$
\ell_{i_1\cdots i_s}(x_1,\ldots,x_s) = a_{i_1}(x_1)\cdots a_{i_s}(x_s),
$$

where

$$
a_j(x) = \begin{cases} 1-x & \text{if } j = 0, \\ x & \text{if } j = 1, \end{cases}
$$

so that $\ell_{i_1\cdots i_s}$ has the value 1 at the vertex $(i_1,\ldots,i_s)$ and the value 0 at all other vertices. Then we find

$$W_{i_1\cdots i_s} = \frac{1}{2^s} - \frac{1}{N}\sum_{j=1}^{N-1}\ell_{i_1\cdots i_s}(\mathbf{x}_j), \qquad i_1,\ldots,i_s = 0 \text{ or } 1. \tag{8.10}$$

This is the formula used in practice to find the vertex weights for rule B. A useful relation, following from the inversion symmetry of the interior quadrature points, is

$$W_{i_1 i_2\cdots i_s} = W_{1-i_1,1-i_2,\cdots,1-i_s}. \tag{8.11}$$

(Note that if $\mathbf{x}$ belongs to the integration lattice so does $-\mathbf{x}$, and therefore so does $(1,1,\ldots,1) - \mathbf{x}$.)

The vertex weights of rule B have an interesting interpretation: equation (8.10) is equivalent to

$$W_{i_1\cdots i_s} - \frac{1}{2^s N} = (I - T)\ell_{i_1\cdots i_s}. \tag{8.12}$$

This follows from the application of rule T given by (8.5) to the function $\ell_{i_1\cdots i_s}$, which gives

$$T\ell_{i_1\cdots i_s} = \frac{1}{2^s N} + \frac{1}{N}\sum_{j=1}^{N-1}\ell_{i_1\cdots i_s}(\mathbf{x}_j).$$

The right-hand side of (8.12) is the error in the symmetric rule T applied to a multilinear function having the value 1 at one vertex and zero at all other vertices (recall Figure 8.1), while the left-hand side is just the departure of an appropriate vertex weight for rule B from the corresponding value for the symmetric rule T. In simple terms, rules B and T are close to each other to the extent that rule T does a good job of integrating multilinear functions.

Later we shall see (in Section 8.4) examples in which some of the weights $W_{i_1\cdots i_s}$ turn out to be negative or very large, even many times larger than $1/N$. In such cases (8.12) tells us that the symmetric rule is doing a very bad job indeed of integrating the basic multilinear functions.

Of particular significance is the vertex variance

$$v_B := v(B) = \sum_{i_s=0}^{1}\cdots\sum_{i_1=0}^{1} W_{i_1\cdots i_s}^2 \tag{8.13}$$

of the optimal rule B. Since the average of the vertex weights $W_{i_1\cdots i_s}$ is $1/(2^s N)$, it follows from (8.13) that

$$v_B = \frac{1}{2^s N^2} + \sum_{i_s=0}^{1}\cdots\sum_{i_1=0}^{1}\left(W_{i_1\cdots i_s} - \frac{1}{2^s N}\right)^2,$$

or

$$v_B = v(T) + \sum_{i_s=0}^{1} \cdots \sum_{i_1=0}^{1} \left((I-T)\ell_{i_1\cdots i_s}\right)^2 , \qquad (8.14)$$

in the last step using (8.6) and (8.12). Thus the vertex variance (or, more precisely, its square root) is a very direct measure of the failure of the symmetric rule to integrate the basic multilinear functions accurately.

8.4 Numerical experiments—the key role of the vertex variance v_B

In this section, which follows Niederreiter and Sloan (1993) closely, we look at the integration of nonperiodic functions by the family of lattice rules

$$Qf := \frac{1}{N} \sum_{j=0}^{N-1} f\left(\left\{\frac{j}{N}\left(1, \ell, \ldots, \ell^{s-1}\right)\right\}\right), \qquad (8.15)$$

where N is prime, and ℓ is an integer satisfying $1 \leqslant \ell \leqslant \lfloor N/2 \rfloor$. Thus Q is a lattice rule of rank 1, in which the vector $\mathbf{z}$ in (4.1) has the classical Korobov form

$$\mathbf{z} = (1, \ell, \ell^2, \ldots, \ell^{s-1}),$$

in which all that remains to be chosen is the single integer ℓ.

How should we decide which of two given rules of the form (8.15) is the better? On the one hand, the considerations in the preceding section, and in particular (8.14), suggest that one should choose the rule with the smaller value of v_B, the vertex variance of the optimal rule B, since v_B can be small only if the symmetric rules do a good job of integrating multilinear functions. On the other hand, we might argue that because the class of nonperiodic integrands includes the periodic integrands as a special case, we should choose the rule with the smaller value of P_α (see (4.13)), with α chosen to be, say, 2. The surprising truth to emerge from numerical experiments is that the two different criteria may select very different rules. In particular, it turns out that a rule with a very good (that is, small) value of P_2 may have an extraordinarily large value of v_B, and that as a consequence selection on the basis of P_2 alone is very dangerous.

In Table 8.1 we show selected rules found by Niederreiter and Sloan (1993) in a complete search over all rules of the form (8.15) with $s = 8$ and $N = 10\,007$. The search was therefore over the $\lfloor N/2 \rfloor = 5003$ nonzero values of ℓ. The complete list was ranked separately with respect to v_B and P_2; the second column shows the rank order with respect to v_B, the third column the rank order with respect to P_2. The vertex variance v_B was computed from (8.13) after computing the 256 individual vertex weights $W_{i_1\cdots i_8}$ using (8.10) and (8.11). Every rule in the complete list of 5003

Table 8.1 Selected eight-dimensional rules of the form (8.15), with $N = 10\,007$. The second column shows the position of the rule in the complete list of 5003 such rules when ranked by increasing values of v_B (the vertex variance of the rule B). The third shows the position when ranked by increasing values of P_2

ℓ	$\#(v_B)$	$\#(P_2)$	$N^2 v_B$	P_2	$N \max |W|$	$N \sum |W|$
1855	1	2677	22.69	10.85	0.83	64.26
1206	59	4981	30.33	28.04	0.99	71.71
3335	4963	1	8715.05	9.15	21.00	1200.84
1	5003	5003	2538472.89	5814.49	1072.30	7556.37

items turns out to contain some negative vertex weights. This is apparent for the particular rules in the table from the fact that in the last column

$$N \sum_{i_s=0}^{1} \cdots \sum_{i_1=0}^{1} |W_{i_1 \cdots i_s}| > N \sum_{i_s=0}^{1} \cdots \sum_{i_1=0}^{1} W_{i_1 \cdots i_s} = 1.$$

Nevertheless, we see from the second-last column that the individual vertex weights in the first two rules in the table never exceed $1/N$ in absolute value, which may seem an acceptable situation if we recall that the weight at each interior point of the lattice rule is $1/N$.

The remarkable feature of Table 8.1 is that the value $\ell = 3335$, which as it happens yields the smallest value of P_2, gives an extraordinarily large value of the vertex variance, and individual vertex weights greater in absolute value than $20/N$. There is as yet little understanding of this dramatic mismatch, but the same phenomenon certainly occurs at other values of s and N, and it may well be a general truth for large s and N that there exist rules with very good values of P_2 but very bad values of the vertex variance v_B. In the light of equation (8.14) such rules must be viewed with great suspicion.

In Tables 8.2 and 8.3 we show the results of the rules in Table 8.1 when applied to the Genz package (Genz 1984, 1987) of test integrals. This package will be explained fully in Chapter 11. For the present it suffices to say that there are six classes of integrands, with the principal characteristic of each class being identified by the descriptor in the table. Each integrand has two kinds of parameter, the first determining the degree of difficulty, and the second the location of peaks, etc. Niederreiter and Sloan (1993) followed Genz's recommendations for the 'difficulty' parameters, except that for the 'corner peak' case Genz's recommended value of $h = 600$ was replaced by $h = 100$, for the very good reason that with the original value of $h = 600$ all four rules found the integrands too difficult to yield useful

Table 8.2 Tests of the rules of Table 8.1 on the Genz package. All rules have $s = 8$ and $N = 10\,007$. Rule Q is of the form (8.15), and rules T and B are defined by (8.5) and (8.8)

| | Median number of correct digits | | | | | |
| | $\ell = 1855$ | | | $\ell = 1206$ | | |
Integrand class	Q	T	B	Q	T	B
Oscillatory	3.49	3.76	4.03	2.94	3.00	3.87
Product peak	3.58	3.63	3.62	3.24	3.24	3.29
Corner peak	2.00	2.89	2.60	1.92	2.66	2.51
Gaussian	2.62	2.62	2.61	2.39	2.39	2.39
Continuous	4.79	4.97	4.78	4.45	4.50	4.53
Discontinuous	2.95	2.87	3.10	3.23	3.19	3.16
Average	3.24	3.46	3.46	3.03	3.16	3.29

information. (No such problem is encountered in Chapter 11, where we carry out nonlinear transformations before applying the lattice rules. Of course if it is known in advance that the main variation of the integrand occurs near a corner or a boundary, then the nonlinear transformation approach has a clear advantage, since it has the effect of concentrating points near the boundaries.)

The results in Tables 8.2 and 8.3 were obtained by taking 20 'random' integrands of the same difficulty for each of the six integrand classes. The entries in the tables show the median number of correct digits for the sample of 20. The last row shows the mean of these median numbers over all six classes.

These results in Tables 8.2 and 8.3 provide striking empirical evidence that the vertex variance v_B does indeed play a key role: in every case the average error increases by one to two orders of magnitude in going from the first to the last rule. Yet by the classical criterion P_2 for periodic integrands the rule used to obtain the first set of results in Table 8.3 would be the best of the four!

There are, of course, other criteria for comparing lattice rules. Perhaps the 'figure of merit' ρ (see (5.3)) would tell a different story? In reality, however, the values of ρ are of no help: the ρ values for the four rules in Table 8.1 are, respectively, 3, 1, 3, and 1. Thus the poorly performing third rule has as good a ρ value as any other, and a much better value than the manifestly superior second rule. So the mystery remains.

Table 8.3 Tests of the rules of Table 8.1 on the Genz package. All rules have $s = 8$ and $N = 10\,007$. Rule Q is of the form (8.15), and rules T and B are defined by (8.5) and (8.8)

| | Median number of correct digits | | | | | |
| | $\ell = 3335$ | | | $\ell = 1$ | | |
Integrand class	Q	T	B	Q	T	B
Oscillatory	1.51	1.51	2.19	0.17	0.17	0.00
Product peak	2.26	2.26	2.49	1.11	1.11	0.00
Corner peak	1.04	0.99	0.68	0.00	0.00	0.00
Gaussian	1.71	1.71	1.77	0.50	0.50	0.00
Continuous	3.30	3.30	3.67	1.83	1.83	0.00
Discontinuous	1.44	1.42	1.44	0.60	0.60	0.00
Average	1.88	1.87	2.04	0.70	0.70	0.00

One other conclusion might be drawn from the tables: as long as the vertex variance is kept small, there is not much difference between the results with the symmetric rule T and the optimal rule B, and so not much advantage, if any, in working with B rather than the simpler rule T, for which no weights need be computed.

Our tentative recommendation is therefore: For the application of lattice rules to nonperiodic functions, choose a rule with a small value of v_B, the vertex variance of the optimal rule B. Once this is done, numerical calculations can be carried out with the much simpler symmetric rule T, or perhaps even with the original lattice rule Q.

8.5   Theoretical properties of the optimal rule B

Theoretical properties of rule B are known from the work of Niederreiter and Sloan (1993). Here we indicate these theoretical results only briefly, not only because the arguments are intricate, but also because the results are in some ways incomplete: in particular, there is as yet no clear understanding of the phenomenon observed in the preceding section, that a rule which is best possible from the point of view of a classical criterion can be very bad when used to integrate nonperiodic integrands.

The 'optimal' property of rule B lies in an error bound established by Niederreiter and Sloan (1993). Suppose that each of the mixed first partial derivatives

$$\frac{\partial^{q_1+\cdots+q_s} f}{\partial x_1^{q_1} \cdots \partial x_s^{q_s}}, \qquad 0 \leqslant q_k \leqslant 1, \qquad 1 \leqslant k \leqslant s,$$

is continuous on C^s. Then it can be shown that the general vertex-modified rule M has as an error bound

$$|Mf - If| \leqslant t(M)V(f),$$

where $t(M)$ is a constructible constant related to the L_2 discrepancy of the rule, and $V(f)$ is given for $s = 2$ by

$$(V(f))^2 = \int_0^1 \left(\frac{\partial f}{\partial x}(x,1)\right)^2 dx + \int_0^1 \left(\frac{\partial f}{\partial y}(1,y)\right)^2 dy$$
$$+ \int_0^1 \int_0^1 \left(\frac{\partial^2 f}{\partial x \partial y}(x,y)\right)^2 dx\, dy,$$

and by an analogous formula for other values of s. The optimal aspect of rule B is that the constant $t(M)$ turns out to be minimized by the choice $M = B$. This fact provides an alternative characterization of rule B. (As a matter of historical record, that is how rule B was first approached, its property of integrating exactly all multilinear functions being observed only later.)

Given our insistence on choosing a rule with small vertex variance when integrating nonperiodic functions, it is fortunate that we know that such rules always exist: Niederreiter and Sloan (1993) show, for any s and N, that there exists a lattice rule Q such that

$$v_B \leqslant c\frac{(\log N)^{2s}}{N^2},$$

with c independent of N. We omit the proof.

8.6   The special case of two dimensions

Although the two-dimensional case may not be of great practical interest, it does have a special charm. In Section 4.3 we discussed briefly the Fibonacci lattice rules

$$Q_k f = \frac{1}{F_k} \sum_{j=0}^{F_k-1} f\left(\left\{\frac{j}{F_k}(1,F_{k-1})\right\}\right), k = 3,4,\ldots,$$

where $F_1, F_2, \ldots$ are the successive values of the Fibonacci sequence $1,1,2,$ $3,5,\ldots$. For application to nonperiodic integrands we also need the corresponding 'symmetric' and 'optimal' vertex-modified rules, defined by (see (8.5) and (8.8))

$$T_k f \;:=\; \frac{1}{4F_k}[f(0,0) + f(1,0) + f(0,1) + f(1,1)]$$

$$+ \frac{1}{F_k} \sum_{j=1}^{F_k-1} f\left(\left\{\frac{j}{F_k}(1, F_{k-1})\right\}\right),$$

and

$$B_k f := \sum_{m=0}^{1} \sum_{n=0}^{1} W_{nm}^{(k)} f(n,m) + \frac{1}{F_k} \sum_{j=1}^{F_k-1} f\left(\left\{\frac{j}{F_k}(1, F_{k-1})\right\}\right),$$

with

$$W_{nm}^{(k)} = \frac{1}{4} - \frac{1}{F_k} \sum_{j=1}^{F_k-1} \ell_{nm}\left(\left\{\frac{j}{F_k}(1, F_{k-1})\right\}\right) \quad \text{for } m, n = 0 \text{ or } 1.$$

For the symmetric Fibonacci lattice rules, Niederreiter and Sloan (in press) proved the following theorem.

Theorem 8.4. *For $k \geqslant 3$ the symmetric Fibonacci rule T_k integrates every bilinear function exactly if and only if k is odd.*

This theorem points to a clear difference between the odd-labelled and even-labelled Fibonacci lattices. We shall see later that this difference is apparent in numerical calculations.

Since the defining property of rule B_k is that it integrates all bilinear functions exactly, equivalent statements of the theorem are:

$$B_k f = T_k f \ \text{ for all continuous } f \text{ if and only if } k \text{ is odd,}$$

and

$$W_{00}^{(k)} = W_{10}^{(k)} = W_{01}^{(k)} = W_{11}^{(k)} = \frac{1}{4F_k} \ \text{ if and only if } k \text{ is odd.}$$

In Table 8.4 we show the weights $W_{nm}^{(k)}$ for the leading Fibonacci rules. The behaviour is as predicted.

The proof of Theorem 8.4 rests upon the following property of the Fibonacci lattices, established by Niederreiter and Sloan (in press, Theorems 3 and 4).

Lemma 8.5. *The lattice L_k corresponding to the Fibonacci rule Q_k is invariant under rotation by $\pi/2$ (and so has a square unit cell) if and only if k is odd.*

The truth of this lemma is apparent to the eye. For example, Figure 8.2 shows the quadrature points for the 34-point Fibonacci rule. Since $34 = F_9$ and 9 is odd, the theorem tells us that we should see in the diagram a square

Table 8.4 Vertex weights for the Fibonacci
rule B_k

k	$N = F_k$	$W_{00}^{(k)} = W_{11}^{(k)}$	$W_{01}^{(k)} = W_{10}^{(k)}$
3	2	$\frac{1}{8}$	$\frac{1}{8}$
4	3	$\frac{11}{108}$	$\frac{7}{108}$
5	5	$\frac{1}{20}$	$\frac{1}{20}$
6	8	$\frac{5}{128}$	$\frac{3}{128}$
7	13	$\frac{1}{52}$	$\frac{1}{52}$
8	21	$\frac{79}{5292}$	$\frac{47}{5292}$
9	34	$\frac{1}{136}$	$\frac{1}{136}$
10	55	$\frac{69}{12\,100}$	$\frac{41}{12\,100}$

unit cell; and we do, one square unit cell being illustrated. On the other
hand, Figure 8.3 shows the quadrature points for the 55-point Fibonacci
rule. In this diagram, because $55 = F_{10}$ and 10 is even, we do not expect
to see any square unit cell, and clearly there is none. The formal proof of
the lemma (which we omit) uses properties of the Fibonacci numbers.

The remaining link in the proof of Theorem 8.4 is the next lemma, again
from Niederreiter and Sloan (in press, Theorem 2).

Lemma 8.6. *Let Q be a two-dimensional lattice rule in which every quadrature point other than $\mathbf{0}$ lies in $(0,1)^2$. If the integration lattice for Q is
invariant under rotation by $\pi/2$, then the corresponding symmetric rule T
is exact for every bilinear function.*

Proof Since rule T is automatically exact for linear functions (see Theorem 8.3), it is sufficient to show that T is exact for the function $g(x,y) =
(x - \frac{1}{2})(y - \frac{1}{2})$. (This is because every bilinear function is a linear combi-
nation of 1, x, y, and $g(x,y)$.) Let R be the operation of rotation of the
unit square about the point $(\frac{1}{2},\frac{1}{2})$ by $\pi/2$ in the positive direction. Then
$Rg = -g$. Rule T is invariant under the rotation R, by the assumption, so
we have $Tg = TRg = -Tg$, giving $Tg = 0 = Ig$. ∎

Niederreiter and Sloan (in press) also give an alternative proof of Lemma
8.6, based on the error expression (8.7).

In Table 8.5 we show numerical results for the particular function

$$f(x,y) = (e - 2)^{-1}ye^{xy},$$

in which the constant factor has been chosen so that $If = 1$. The column giving the numerical approximations for the symmetric rule T_k shows

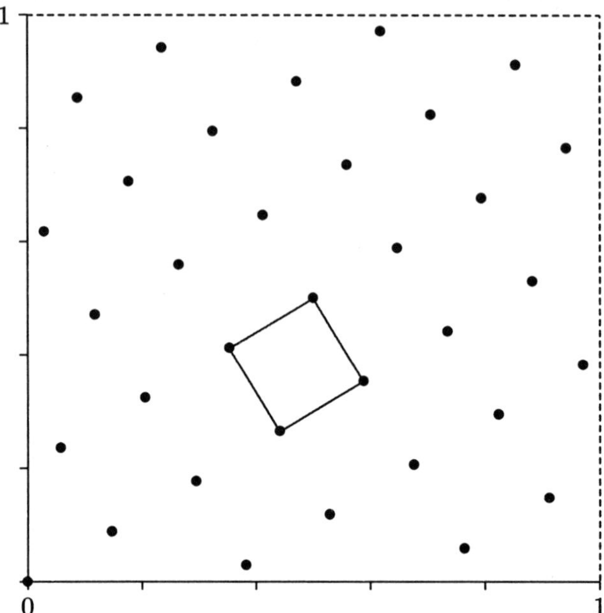

Fig. 8.2 The quadrature points for the Fibonacci lattice rule with $N = 34$.

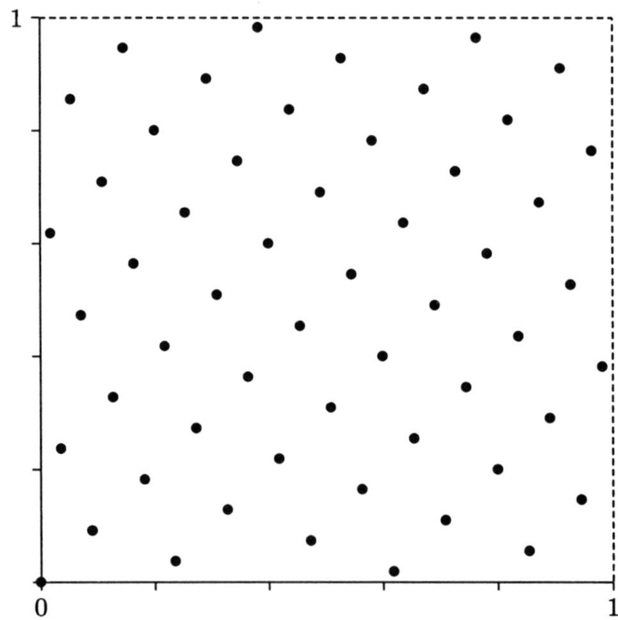

Fig. 8.3 The quadrature points for the Fibonacci lattice rule with $N = 55$.

Table 8.5 Numerical results for the function $f(x,y) = (e-2)^{-1}ye^{xy}$

k	$N = F_k$	Error in Q_k	Error in T_k	Error in B_k
5	5	0.23794(0)	0.20891(−1)	0.20891(−1)
6	8	0.16975(0)	0.79809(−2)	0.10708(−1)
7	13	0.95328(−1)	0.42228(−2)	0.42228(−2)
8	21	0.66648(−1)	0.50211(−2)	0.22116(−2)
9	34	0.37200(−1)	0.86366(−3)	0.86366(−3)
10	55	0.25788(−1)	0.22581(−2)	0.50976(−3)
11	89	0.14348(−1)	0.19302(−3)	0.19302(−3)
12	144	0.99083(−2)	0.92109(−3)	0.13643(−3)
13	233	0.55045(−2)	0.49854(−4)	0.49854(−4)
14	377	0.37945(−2)	0.36176(−3)	0.42194(−4)
15	610	0.21066(−2)	0.14997(−4)	0.14997(−4)
16	987	0.14510(−2)	0.13984(−3)	0.14461(−4)
17	1597	0.80531(−3)	0.50566(−5)	0.50566(−5)
18	2584	0.55452(−3)	0.53685(−4)	0.52511(−5)

very clearly an alternation in accuracy between the odd-labelled and even-labelled Fibonacci rules, as might be expected from Theorem 8.4. The 'optimal' rule B_k consistently gives the best results. Both of the rules T_k and B_k are markedly superior to Q_k.

The results in this section have no obvious extension to higher dimensions, because explicit constructions analogous to the Fibonacci lattice are not known in higher dimensions.

8.7   Is nonlinear transformation a better option?

We end this chapter with some common-sense advice about when to use, and when not to use, the direct methods for nonperiodic functions introduced in this chapter.

The chief deficiency of the direct methods lies in their modest rate of convergence when compared to the nonlinear transformation approach (see Section 2.12). The rule Q of (8.1) applied directly to a nonperiodic integrand f clearly cannot be expected to have a convergence rate better than $O(N^{-1})$, since Qf generally differs from the symmetric rule Tf of (8.5) by a term of order N^{-1}—and in any case even in the one-dimensional setting this (rectangle) rule is only an $O(N^{-1})$ rule. We believe that even the

Table 8.6 Numerical results for the function $f(x,y) = (e-2)^{-1}ye^{xy}$

k	$N = F_k$	Error in Q_k using (2.30)	Error in Q_k using (2.32)
5	5	0.11043(0)	0.50162(−1)
6	8	0.62375(−1)	0.31257(−1)
7	13	0.33491(−1)	0.23111(−2)
8	21	0.15904(−1)	0.82941(−3)
9	34	0.73139(−2)	0.28846(−4)
10	55	0.32482(−2)	0.46728(−4)
11	89	0.14180(−2)	0.15875(−5)
12	144	0.60829(−3)	0.26025(−5)
13	233	0.25805(−3)	0.88336(−7)
14	377	0.10832(−3)	0.14504(−6)
15	610	0.45114(−4)	0.49214(−8)
16	987	0.18657(−4)	0.80829(−8)
17	1597	0.76712(−5)	0.27425(−9)
18	2584	0.31381(−5)	0.45045(−9)

symmetric rule T will not usually yield an error rate better than $O(N^{-1})$; for example, the results for the Fibonacci lattice rules with even values of k in Table 8.5 seem to be showing convergence of this order. At this time, it is not known whether the 'optimal' rule B can yield a better convergence rate than $O(N^{-1})$ if the lattice rule is well chosen, but any gain is likely to be modest.

In contrast, the theory of the foregoing chapters tells us that, provided f is sufficiently well behaved, the nonlinear transformation (as used in (2.28) and (2.29)) can achieve much better asymptotic rates than $O(N^{-1})$. For example, the simplest choice (2.30) of the transformation function leads, via the error bound (4.9) and Theorem 4.2, to an error bound of order $O((\log N)^{2s}N^{-2})$ for a well-chosen lattice rule of rank 1. The higher-order transformation of (2.31) or (2.32) leads by the same path to an error bound of order $O((\log N)^{4s}N^{-4})$. That the same rate of convergence can also be achieved by 2^s copies of well-chosen rank-1 rules follows from Theorem 6.9.

Table 8.6 illustrates this by showing results obtained by applying the nonlinear transformations (2.30) and (2.32) to the same example as in Table 8.5 and then using the Fibonacci rules; in the latter case the improvement over the result B_k of the direct approach is dramatic. (Of course we ought not to forget that in low-dimensional cases there are still better nonlattice options; for the present example a product of five-point Gauss rules (requiring just 25 function evaluations) has an error of approximately 9.84×10^{-13}!)

If the integrand f is not smooth but has singularities on the boundaries then the nonlinear transformation is beneficial in weakening the singularities. Indeed, empirical evidence (Beckers and Haegemans 1992a) suggests that in this situation the strength of the nonlinear transformation should be increased even beyond (2.32).

Order of convergence, however, is not the only issue. Other relevant questions are: how large does N need to be before the asymptotic order of convergence is seen numerically? And how large is the constant that stands in front of the leading order? If a rough estimate of an integral is required with a relatively small number of function evaluations, or if the integrand is not smooth in the interior of the unit cube, then the direct approach may come into its own. The main competition now is perhaps with Monte Carlo and quasi-Monte Carlo methods (see Sections 1.5 and 1.6). To see what can happen with transformation methods when N is small, it is useful to look again at the example in Tables 8.5 and 8.6, now concentrating in the latter case on the simpler nonlinear transformation (2.30). We see that for all values of N up to 1597 the error from the optimal rule B_k is smaller than that from the transformation method, and that not until the very last row is the transformation result the better one. In the case of more challenging higher-dimensional integrals there is numerical evidence that nonlinear transformations, especially strong ones, can give very poor results if the integral is very difficult and the number of points small.

9

Lattice rules—other topics

9.1 The number of lattice rules

In this section we seek an explicit expression for the number of distinct lattice rules of a given order N. Note that rules that are geometrically equivalent (in the sense of Section 2.13) but not identical are counted separately here. It would make sense to ask also for the number of geometrically inequivalent rules of a given order N, but no expression for this (smaller) number is yet known.

We recall from Definition 2.6 that every integration lattice has a dual lattice. Moreover, there exists a generator matrix for the dual lattice. In other words, with the generator matrix denoted by B', the dual lattice consists of all the integer linear combinations of the rows of B'. The entries of the matrix B' are integers.

It is obvious that the dual lattice is unchanged by the following row operations on its generator matrix:

 (i) multiply a row by -1,
 (ii) interchange two rows,
(iii) add an integer multiple of one row to another row.

If the new generator matrix resulting from a sequence of such operations is denoted by B, then B is related to B' by $B = UB'$, where U is an integer unimodular matrix (that is, it has determinant ± 1). Since $|\det B'| = N$, it follows that $|\det B| = N$. By applying a well-chosen sequence of such operations, Lyness and Sørevik (1989) show that it is possible to obtain a generator matrix B whose elements b_{ij} satisfy $b_{ij} = 0$ for $i > j$, and also

$$b_{jj} > 0 \text{ for } 1 \leqslant j \leqslant s, \quad \prod_{j=1}^{s} b_{jj} = N, \text{ and } 0 \leqslant b_{ij} < b_{jj} \text{ for } 1 \leqslant i < j \leqslant s.$$

$$(9.1)$$

Thus B is an upper triangular matrix having positive diagonal elements, with all the elements in each column being nonnegative and less than the diagonal element. In the language of integer matrices (for example, see Newman 1972), such a B is said to be in 'Hermite normal form'. The argument in Lyness and Sørevik (1989) shows that the generator matrix for a dual lattice may be expressed uniquely in Hermite normal form. Conversely, each matrix in Hermite normal form is a generator matrix for the dual of some integration lattice.

It follows that the total number of distinct s-dimensional lattice rules having order N equals the number $T_s(N)$ of $s \times s$ matrices in Hermite normal form having determinant N. It can be shown that $T_s(\cdot)$ is a multiplicative function, that is,

$$T_s(MN) = T_s(M)T_s(N), \tag{9.2}$$

whenever M and N are relatively prime. From this it follows that $T_s(N)$ may be found for any N if we know its value for prime powers p^β. The required result for prime powers is given in the following theorem, whose proof may be found in Lyness and Sørevik (1989) or Newman (1972).

Theorem 9.1. *For p prime and β a positive integer, we have*

$$T_s(p^\beta) = \prod_{i=1}^{s-1} \left[\frac{p^{\beta+i} - 1}{p^i - 1} \right] = \prod_{i=1}^{\beta} \left[\frac{p^{s+i-1} - 1}{p^i - 1} \right].$$

A table of values of $T_s(N)$ for different values of s and N may be found in Lyness and Sørevik (1989). Even with small values of N, the value $T_s(N)$ grows very rapidly as s increases. For example, if $N = 256 = 2^8$ and $s = 7$, we find that $T_7(N) \approx 9.56 \times 10^{14}$. From Lyness and Sørevik (1989), there exists $\kappa(s)$ such that

$$\frac{N^s - 1}{N - 1} \leqslant T_s(N) \leqslant \kappa(s)N^{s-1} \log \log N,$$

with the lower bound being achieved when N is prime. Valid values for $\kappa(s)$ include $\kappa(3) = 2.93$ and $\kappa(4) = 3.53$.

9.2   The number of lattice rules having given invariants

We recall that the rank and invariants of any lattice rule are uniquely determined. If a lattice rule of order N has rank r, then its r invariants $n_1, \ldots, n_r$ satisfy

$$n_r > 1, \text{ and } n_{k+1} \text{ divides } n_k \text{ for } 1 \leqslant k \leqslant r - 1, \tag{9.3}$$

and

$$N = n_1 n_2 \cdots n_r. \tag{9.4}$$

One simple consequence of (9.3) and (9.4) is that N contains n_r^r as a factor. Thus if N is a prime number or a product of different prime numbers then the lattice rule must be of rank 1. In contrast, if N is neither prime nor the product of distinct primes one can obtain different sets of invariants that satisfy (9.3) and (9.4). For instance, if $s \geqslant 2$ and $N = 36 = 2^2 3^2$, there may exist rank-1 rules with $n_1 = 36$, and rank-2 rules with invariants $n_1 = 18$, $n_2 = 2$, or with $n_1 = 12$, $n_2 = 3$, or $n_1 = n_2 = 6$.

Although Theorem 9.1 combined with (9.2) gives the total number of distinct lattice rules of order N, it does not give the number of lattice rules having a given set of invariants $n_1, \ldots, n_r$, that is, the number of lattice rules of order $n_1 n_2 \cdots n_r$ having the form

$$\frac{1}{n_1 n_2 \cdots n_r} \sum_{j_r=0}^{n_r-1} \cdots \sum_{j_1=0}^{n_1-1} f\left(\left\{\frac{j_1}{n_1}\mathbf{z}_1 + \cdots + \frac{j_r}{n_r}\mathbf{z}_r\right\}\right), \qquad (9.5)$$

with $n_1, n_2, \ldots, n_r$ satisfying (9.3). We shall use $\nu_s(n_1, \ldots, n_r)$ to denote this quantity. The task in this section is to find an explicit expression for it. Without loss of generality, we shall always assume in this section that the integer vectors $\mathbf{z}_k$ in (9.5) have components z_{ki} satisfying

$$0 \leqslant z_{ki} \leqslant n_k - 1, \quad 1 \leqslant i \leqslant s, \quad 1 \leqslant k \leqslant r. \qquad (9.6)$$

Suppose N has the prime factorization $p_1^{\gamma_1} p_2^{\gamma_2} \cdots p_q^{\gamma_q}$. For convenience let us assume that $\gamma_1 = \max_{1 \leqslant i \leqslant q} \gamma_i$. Then for any r satisfying $1 \leqslant r \leqslant \min(s, \gamma_1)$ one can always obtain a set of r invariants that satisfy (9.3) and (9.4) by setting $n_1 = p_1^{\gamma_1 - r + 1} p_2^{\gamma_2} \cdots p_q^{\gamma_q}$, $n_2 = \cdots = n_r = p_1$. In general, any set of invariants $n_1, \ldots, n_r$ satisfying (9.3) and (9.4) must be of the form

$$n_k = p_1^{\gamma_{1k}} p_2^{\gamma_{2k}} \cdots p_q^{\gamma_{qk}}, \quad 1 \leqslant k \leqslant r, \qquad (9.7)$$

where

$$0 \leqslant \gamma_{ir} \leqslant \gamma_{i,r-1} \leqslant \cdots \leqslant \gamma_{i1} \text{ and } \sum_{k=1}^{r} \gamma_{ik} = \gamma_i, \ 1 \leqslant i \leqslant q. \qquad (9.8)$$

As a first step in finding an expression for $\nu_s(n_1, \ldots, n_r)$, the next result shows that $\nu_s(n_1, \ldots, n_r)$ is in a certain sense a multiplicative function in each of its arguments.

Theorem 9.2. *Suppose $n_1, \ldots, n_r$ are given invariants expressed in the form (9.7) and (9.8), so that $n_1 n_2 \cdots n_r = p_1^{\gamma_1} p_2^{\gamma_2} \cdots p_q^{\gamma_q}$. Then*

$$\nu_s(n_1, \ldots, n_r) = \prod_{i=1}^{q} \nu_s(p_i^{\gamma_{i1}}, \ldots, p_i^{\gamma_{it_i}}),$$

where t_i, for $1 \leqslant i \leqslant q$, is the largest integer t^* for which $\gamma_{it^*} > 0$, with each t_i satisfying $1 \leqslant t_i \leqslant r$.

Proof Let $\mathcal{A}(Q)$ be the set of quadrature points for any lattice rule Q having invariants $n_1, \ldots, n_r$. Considered as a group, it follows from (3.5) that $\mathcal{A}(Q)$ may be expressed as the direct sum of q subgroups,

$$\mathcal{A}(Q) = \mathcal{S}_1(Q) \oplus \cdots \oplus \mathcal{S}_q(Q),$$

where $\mathcal{S}_i(Q)$ is of order $p_i^{\gamma_i}$, $1 \leqslant i \leqslant q$. These groups $\mathcal{S}_i(Q)$, $1 \leqslant i \leqslant q$, are the Sylow p_i subgroups of $\mathcal{A}(Q)$.

We want the number of distinct lattice rules having invariants $n_1, \ldots, n_r$. Clearly, the $p_i^{\gamma_i}$ elements of $\mathcal{S}_i(Q)$ are the points of a lattice rule having invariants $p_i^{\gamma_{i1}}, \ldots, p_i^{\gamma_{it_i}}$. (See, for example, Theorem 3.29.) Hence for $1 \leqslant i \leqslant q$, there are $\nu_s(p_i^{\gamma_{i1}}, \ldots, p_i^{\gamma_{it_i}})$ distinct Sylow p_i subgroups. Since the elements of $\mathcal{S}_i(Q)$ consist of those elements of $\mathcal{A}(Q)$ whose order is a power of p_i, the q Sylow p_i subgroups for a given lattice rule having invariants $n_1, \ldots, n_r$ are unique. Hence we conclude that the number of distinct lattice rules having invariants $n_1, \ldots, n_r$ is

$$\nu_s(n_1, \ldots, n_r) = \prod_{i=1}^{q} \nu_s(p_i^{\gamma_{i1}}, \ldots, p_i^{\gamma_{it_i}}),$$

as claimed. ∎

This last theorem shows that we need only consider $\nu_s\left(p^{\beta_1}, \ldots, p^{\beta_r}\right)$, where p is prime, and $\beta_1 \geqslant \beta_2 \geqslant \cdots \geqslant \beta_r > 0$. The first and simplest such case to look at is the number of rank-1 lattice rules having order p^β.

Theorem 9.3. *For p a prime number and β a positive integer, the number of distinct rank-1 lattice rules of order p^β is*

$$\nu_s\left(p^\beta\right) = \frac{p^{\beta s} - p^{(\beta-1)s}}{p^\beta - p^{\beta-1}}.$$

Proof We see from (9.5) that a rank-1 lattice rule of order p^β may be written as

$$Q(\mathbf{z}, p^\beta) = \frac{1}{p^\beta} \sum_{j=0}^{p^\beta - 1} f\left(\left\{\frac{j}{p^\beta}\mathbf{z}\right\}\right).$$

Without loss of generality we may assume that the components of $\mathbf{z}$ satisfy $0 \leqslant z_i \leqslant p^\beta - 1$, $1 \leqslant i \leqslant s$. As in Chapter 3, the set of p^β points for this lattice rule forms a group, with the group operation given by (3.4). In this case the group is a cyclic group generated by the element $\mathbf{c} = \mathbf{z}/p^\beta$. Since this generator must have order p^β (that is, p^β is the least positive integer u for which $\{u\mathbf{c}\} = \mathbf{0}$), it follows that at least one of the components of $\mathbf{z}$ is not divisible by p. (Otherwise, if $\mathbf{z} = p\mathbf{z}'$, then $\mathbf{c} = \mathbf{z}/p^\beta = \mathbf{z}'/p^{\beta-1}$ has order at most $p^{\beta-1}$.)

Because $p^\beta - \varphi\left(p^\beta\right) = p^{\beta-1}$ (Ledermann 1964, p.17), where φ is Euler's totient function, we see that though there are $p^{\beta s}$ possible choices of $\mathbf{z}$, $\left(p^{\beta-1}\right)^s$ of them do not produce a rank-1 rule of order p^β since all their components are divisible by p. Thus the apparent total possible number of rank-1 lattice rules of order p^β is given by

$$p^{\beta s} - p^{(\beta-1)s}.$$

However, different choices of $\mathbf{z}$ may produce the same rank-1 lattice rule. The theory of cyclic groups (for example, see Ledermann 1964, p.159) shows that the generating element of a cyclic group of order p^β may be chosen in $\varphi\left(p^\beta\right)$ ways. Thus we obtain

$$\nu_s\left(p^\beta\right) = \frac{p^{\beta s} - p^{(\beta-1)s}}{p^\beta - p^{\beta-1}},$$

which completes the proof.   ∎

We now give an expression for $\nu_s\left(p^{\beta_1},\ldots,p^{\beta_r}\right)$ (Joe and Hunt 1992). An equivalent result has been obtained by Lyness (1993).

Theorem 9.4. *Suppose $p^{\beta_1},\ldots,p^{\beta_r}$ are given invariants such that d of them are distinct. Let $p^{\beta_1^*},\ldots,p^{\beta_d^*}$ be these distinct invariants, with $\beta_1^* > \cdots > \beta_d^* > 0$, and let ω_k be the number of invariants equal to $p^{\beta_k^*}$, $1 \leqslant k \leqslant d$. Defining*

$$\delta_k := \beta_k^* \sum_{j=1}^{k} \omega_j + \sum_{j=k+1}^{d} \omega_j \beta_j^*, \quad 1 \leqslant k \leqslant d,$$

the number of distinct s-dimensional lattice rules having the given invariants is

$$\nu_s\left(p^{\beta_1},\ldots,p^{\beta_r}\right) = \frac{\displaystyle\prod_{k=1}^{r} \left(p^{\beta_k s} - p^{(\beta_k-1)s+k-1}\right)}{\displaystyle\prod_{k=1}^{d}\left[\prod_{j=1}^{\omega_k}\left(p^{\delta_k} - p^{\delta_k-\omega_k+j-1}\right)\right]}. \tag{9.9}$$

In Joe and Hunt (1992) the result was obtained from an argument based on the group-theoretic structure of lattice rules, whereas Lyness (1993) used an argument based on a matrix representation for lattice rules. The proof to be given here is based on the proof of Joe and Hunt (1992), but has been recast from its original group-theoretic terminology (such as is used in the proof of the previous theorem) into lattice rule terminology.

The proof of Theorem 9.4 depends on the following three lemmas. We shall say that a lattice rule Q' is 'embedded' in the lattice rule Q if the quadrature points of Q' are also quadrature points of Q.

Lemma 9.5. *Let*

$$Q_{r-1}f \;=\; \frac{1}{p^{\beta_1}p^{\beta_2}\cdots p^{\beta_{r-1}}} \sum_{j_{r-1}=0}^{p^{\beta_{r-1}}-1} \cdots \sum_{j_1=0}^{p^{\beta_1}-1}$$
$$\times\, f\left(\left\{\frac{j_1}{p^{\beta_1}}\mathbf{z}_1 + \cdots + \frac{j_{r-1}}{p^{\beta_{r-1}}}\mathbf{z}_{r-1}\right\}\right) \qquad (9.10)$$

be a lattice rule having rank $r-1$ and invariants $p^{\beta_1},\ldots,p^{\beta_{r-1}}$.

(i) *Suppose*

$$Q'f = \sum_{j=0}^{p^{\beta_r}-1} f\left(\left\{\frac{j}{p^{\beta_r}}\mathbf{z}_r\right\}\right)$$

is a rank-1 rule with order p^{β_r}, $0 < \beta_r \leqslant \beta_{r-1}$. Then a necessary and sufficient condition for

$$\frac{1}{p^{\beta_1}p^{\beta_2}\cdots p^{\beta_r}} \sum_{j_r=0}^{p^{\beta_r}-1} \sum_{j_{r-1}=0}^{p^{\beta_{r-1}}-1} \cdots \sum_{j_1=0}^{p^{\beta_1}-1}$$
$$\times\, f\left(\left\{\frac{j_1}{p^{\beta_1}}\mathbf{z}_1 + \cdots + \frac{j_{r-1}}{p^{\beta_{r-1}}}\mathbf{z}_{r-1} + \frac{j_r}{p^{\beta_r}}\mathbf{z}_r\right\}\right) \qquad (9.11)$$

to be a lattice rule having rank r and invariants $p^{\beta_1},\ldots,p^{\beta_r}$ is that every rank-1 rule of order p embedded in Q' is not embedded in Q_{r-1}.

(ii) *There are $(p^{r-1}-1)/(p-1)$ distinct rank-1 lattice rules of order p embedded in Q_{r-1}.*

Proof We first assert that a necessary and sufficient condition for (9.11) to be a rank-r rule having the required invariants is that Q_{r-1} and Q' do not have any quadrature points (except the trivial point $\mathbf{0}$) in common, so that (9.11) is a direct sum. The necessity is clear. To prove sufficiency, it is enough to observe that (9.11) is then a rule with $p^{\beta_1}p^{\beta_2}\cdots p^{\beta_r}$ distinct points expressed in canonical form, so revealing the rank to be r and the invariants to be $p^{\beta_1},\ldots,p^{\beta_r}$.

Obviously, if Q_{r-1} and Q' do not have a nontrivial common point, then a rank-1 rule of order p embedded in Q' cannot also be embedded in Q_{r-1}. Also, if every rank-1 rule of order p embedded in Q' is not embedded in Q_{r-1}, then Q_{r-1} and Q' cannot have any nontrivial point in common. To see this, suppose $\mathbf{c} \neq \mathbf{0}$ was such a common point. Then there would exist an integer γ satisfying $0 \leqslant \gamma \leqslant \beta_r - 1$ such that $\{p^\gamma \mathbf{c}\} = \{\mathbf{z}/p\}$, where $\mathbf{z} \in \mathbb{Z}^s$ has at least one component not divisible by p. In this situation, the points $\mathbf{0}, \{\mathbf{z}/p\}, \{2\mathbf{z}/p\},\ldots,\{(p-1)\mathbf{z}/p\}$, which are the points of a rank-1 rule of order p, would belong to both Q_{r-1} and Q', contradicting the assumption. Hence part (i) is proved.

To prove part (ii), for $1 \leqslant k \leqslant r-1$, let h_k take integer values from 0 to $p-1$. Then by setting $j_k = h_k p^{\beta_k - 1}$ in (9.10), we see that Q_{r-1} contains the quadrature points

$$\left\{ \frac{1}{p} (h_1 \mathbf{z}_1 + \cdots + h_{r-1} \mathbf{z}_{r-1}) \right\}, \quad 0 \leqslant h_k \leqslant p-1, \quad 1 \leqslant k \leqslant r-1.$$

There are $p^{r-1} - 1$ such points if we exclude the case $h_1 = \cdots = h_{r-1} = 0$. We shall use $\mathcal{A}$ to denote the set of these points. The points are distinct and different from $\mathbf{0}$, since by assumption all the quadrature points in (9.10) are distinct. Thus the elements of $\mathcal{A}$ are of the form $\{\mathbf{z}/p\}$, where $\mathbf{z} \in \mathbb{Z}^s$ has at least one component not divisible by p. It is easily seen that no other quadrature points of (9.10) are of this form. The set $\mathcal{A} \cup \{\mathbf{0}\}$ is a subgroup of the group of quadrature points under the operation (3.4). Thus for any $\mathbf{c} \in \mathcal{A}$, the points $\mathbf{c}, \{2\mathbf{c}\}, \ldots, \{(p-1)\mathbf{c}\}$ must all be points of $\mathcal{A}$. But these $p-1$ points together with $\mathbf{0}$ are the points of a rank-1 rule of order p embedded in Q_{r-1}. It follows that the $p^{r-1} - 1$ points in $\mathcal{A}$ must form $(p^{r-1}-1)/(p-1)$ distinct rank-1 rules of order p embedded in Q_{r-1}.   ■

Now it is convenient to let $\tilde{\nu}(p^{\beta_1}, \ldots, p^{\beta_r})$ denote the total number of possible formulas (9.11) in which the vectors $\mathbf{z}_k$ satisfy (9.6) with $n_k = p^{\beta_k}$, $1 \leqslant k \leqslant r$, and which correspond to lattice rules having invariants $p^{\beta_1}, \ldots, p^{\beta_r}$. The quantity $\tilde{\nu}(p^{\beta_1}, \ldots, p^{\beta_r})$ is not the same as $\nu(p^{\beta_1}, \ldots, p^{\beta_r})$ since, as we shall see shortly, different collections of the $\mathbf{z}_1, \ldots, \mathbf{z}_r$ may give rise to the same lattice rule. We have already seen from the proof of Theorem 9.3 that

$$\tilde{\nu}(p^{\beta_1}) = p^{\beta_1 s} - p^{(\beta_1 - 1)s}. \tag{9.12}$$

Lemma 9.6. *Each one of the $\nu_s(p)$ distinct rank-1 lattice rules of prime order p is embedded in exactly $(p-1)p^{(\gamma-1)s}$ of the $\tilde{\nu}_s(p^\gamma)$ rank-1 formulas of order p^γ, $\gamma > 0$.*

Proof Let

$$\frac{1}{p} \sum_{j=0}^{p-1} f\left(\left\{\frac{j}{p}\mathbf{z}'\right\}\right)$$

be any one of the $\nu_s(p)$ distinct rank-1 rules of order p. Also let

$$\mathbf{z} = \mathbf{z}' + p\mathbf{h},$$

where $\mathbf{h}$ is an integer vector whose components satisfy

$$0 \leqslant h_i \leqslant p^{\gamma-1} - 1, \quad 1 \leqslant i \leqslant s. \tag{9.13}$$

Since at least one of the components of $\mathbf{z}'$ is not divisible by p, at least one of the components of $\mathbf{z}$ is not divisible by p. Hence

$$\frac{1}{p^\gamma} \sum_{j=0}^{p^\gamma - 1} f\left(\left\{\frac{j}{p^\gamma}\mathbf{z}\right\}\right)$$

is a rank-1 rule of order p^γ. By letting j take on the values $j = kp^{\gamma-1}$ for $0 \leqslant k \leqslant p-1$, it is not hard to see that the original rank-1 rule of order p is embedded in this rank-1 rule of order p^γ. Note also that it follows from (9.13) that the components of $\mathbf{z}$ satisfy $0 \leqslant z_i \leqslant p^\gamma - 1$ if the components of $\mathbf{z}'$ satisfy $0 \leqslant z_i' \leqslant p-1$, $1 \leqslant i \leqslant s$.

The vector $\mathbf{z}'$ for the chosen rank-1 rule of order p may be chosen in $\varphi(p) = p - 1$ ways. Also, (9.13) shows that there are $p^{(\gamma-1)s}$ possibilities for $\mathbf{h}$. Thus the rank-1 rule of order p is embedded in *at least* $(p-1)p^{(\gamma-1)s}$ of the $\tilde{\nu}_s(p^\gamma)$ rank-1 formulas of order p^γ.

To show that the number is exactly $(p-1)p^{(\gamma-1)s}$, we first note from Lemma 9.5(ii) (with $r = 2$) that any rank-1 rule of order p^γ has only one rank-1 rule of order p embedded in it. Also, Theorem 9.3 yields

$$\nu_s(p) \times (p-1)p^{(\gamma-1)s} = (p^s - 1)p^{(\gamma-1)s} = p^{\gamma s} - p^{(\gamma-1)s}.$$

However, this last quantity is just $\tilde{\nu}_s(p^\gamma)$. Hence we see that the exact number is as given. $\blacksquare$

The last lemma we need concerns the number of choices of the integer vectors $\mathbf{z}_1, \ldots, \mathbf{z}_r$ in (9.11) that give rise to the same lattice rule having invariants $p^{\beta_1}, \ldots, p^{\beta_r}$. The group-theoretic arguments required for the proof of this lemma are quite technical, so we shall omit the proof. We remark that the corresponding result found in Joe and Hunt (1992) is much more general than the result given here.

Lemma 9.7. *Suppose we have the r invariants $p^{\beta_1}, \ldots, p^{\beta_r}$. Also, suppose d of the $\beta_1, \ldots, \beta_r$ are distinct so that we have the distinct values $\beta_1^*, \ldots, \beta_d^*$, where $\beta_1^* > \cdots > \beta_d^* > 0$. If ω_k is the number of the $\beta_1, \ldots, \beta_r$ equal to β_k^*, $1 \leqslant k \leqslant d$, and*

$$\delta_k = \beta_k^* \sum_{j=1}^{k} \omega_j + \sum_{j=k+1}^{d} \omega_j \beta_j^*, \quad 1 \leqslant k \leqslant d,$$

then the number of choices of the $\mathbf{z}_1, \ldots, \mathbf{z}_r$ in (9.11) which give rise to the same lattice rule is given by $\kappa\left(p^{\beta_1}, \ldots, p^{\beta_r}\right)$, where

$$\kappa\left(p^{\beta_1}, \ldots, p^{\beta_r}\right) = \prod_{k=1}^{d} \left[\prod_{j=1}^{\omega_k} \left(p^{\delta_k} - p^{\delta_k - \omega_k + j - 1}\right) \right]. \tag{9.14}$$

We are now ready to give the proof of Theorem 9.4

Proof of Theorem 9.4 We shall first use induction to prove that

$$\tilde{\nu}_s\left(p^{\beta_1}, \ldots, p^{\beta_r}\right) = \prod_{k=1}^{r} \left(p^{\beta_k s} - p^{(\beta_k - 1)s + k - 1}\right). \tag{9.15}$$

For $r = 1$ the right-hand side of (9.15) yields $p^{\beta_1 s} - p^{(\beta_1 - 1)s}$, which we know from (9.12) is $\tilde{\nu}_s\left(p^{\beta_1}\right)$. Thus (9.15) is true for $r = 1$.

Let us now assume that

$$\tilde{\nu}_s\left(p^{\beta_1},\ldots,p^{\beta_{r-1}}\right) = \prod_{k=1}^{r-1}\left(p^{\beta_k s} - p^{(\beta_k-1)s+k-1}\right).$$

Also, suppose Q_{r-1} is a rank $r-1$ lattice rule having given invariants $p^{\beta_1},\ldots,p^{\beta_{r-1}}$. We see from Lemma 9.5(i) that a rank-r rule having invariants $p^{\beta_1},\ldots,p^{\beta_{r-1}},p^{\beta_r}$ is obtained from Q_{r-1} by taking the direct sum with a rank-1 lattice rule of order p^{β_r}. However, Lemma 9.5 also shows that the rank-1 rule of order p embedded in such a rule must not be one of the $(p^{r-1}-1)/(p-1)$ rank-1 lattice rules of order p that are embedded in Q_{r-1}.

We see from Lemma 9.6 that each one of these rank-1 rules of order p is embedded in $(p-1)p^{(\beta_r-1)s}$ rank-1 formulas of order p^{β_r}. Thus of the $\tilde{\nu}_s\left(p^{\beta_r}\right) = p^{\beta_r s} - p^{(\beta_r-1)s}$ rank-1 formulas of order p^{β_r}, there are

$$\left((p^{r-1}-1)/(p-1)\right) \times (p-1)p^{(\beta_r-1)s} = (p^{r-1}-1)p^{(\beta_r-1)s}$$

that are *not* permissible. Hence we see that

$$\begin{aligned}
\tilde{\nu}_s\left(p^{\beta_1},\ldots,p^{\beta_r}\right) &= \tilde{\nu}_s\left(p^{\beta_1},\ldots,p^{\beta_{r-1}}\right)\\
&\quad \times \left(p^{\beta_r s} - p^{(\beta_r-1)s} - (p^{r-1}-1)p^{(\beta_r-1)s}\right)\\
&= \prod_{k=1}^{r}\left(p^{\beta_k s} - p^{(\beta_k-1)s+k-1}\right).
\end{aligned}$$

Thus (9.15) holds by induction.

So now we have an expression for the total number of formulas having the invariants $p^{\beta_1},\ldots,p^{\beta_r}$. However, they do not all correspond to distinct lattice rules. We see from Lemma 9.7 that there are $\kappa\left(p^{\beta_1},\ldots,p^{\beta_r}\right)$ ways of choosing the $\mathbf{z}_1,\ldots,\mathbf{z}_r$ in (9.11) to produce the same lattice rule. Using the expression for $\kappa\left(p^{\beta_1},\ldots,p^{\beta_r}\right)$ given in (9.14), we then obtain

$$\nu_s\left(p^{\beta_1},\ldots,p^{\beta_r}\right) = \frac{\tilde{\nu}_s\left(p^{\beta_1},\ldots,p^{\beta_r}\right)}{\kappa\left(p^{\beta_1},\ldots,p^{\beta_r}\right)} = \frac{\displaystyle\prod_{k=1}^{r}\left(p^{\beta_k s} - p^{(\beta_k-1)s+k-1}\right)}{\displaystyle\prod_{k=1}^{d}\left[\prod_{j=1}^{\omega_k}\left(p^{\delta_k} - p^{\delta_k-\omega_k+j-1}\right)\right]}.$$

This completes the proof. ∎

We remark that (9.9) yields $\nu_s\left(p^{\beta_1},\ldots,p^{\beta_r}\right) = 0$ when $r > s$, a sensible conclusion since we already know that the rank of a lattice rule cannot exceed s.

Example 9.8. *Suppose all the* β_k, $1 \leqslant k \leqslant r$, *are equal to* 1 *so that* $d = 1$, $\omega_1 = r$, *and* $\delta_1 = r$. *Then by using* (9.9) *and a few lines of algebra, it is easy to verify that*

$$\underbrace{\nu_s(p,\ldots,p)}_{r \text{ times}} = \prod_{k=1}^{r} \frac{p^{s-r+k} - 1}{p^k - 1}.$$

9.3 Lattice rules for integration over $\mathbb{R}^s$

So far, we have considered the use of lattice rules for integrating functions over the s-dimensional cube. In this section we consider the use of lattice rules for multiple integrals over the whole of $\mathbb{R}^s$, of the form

$$If = \int_{\mathbb{R}^s} f(\mathbf{x})\,d\mathbf{x}. \tag{9.16}$$

Here f is assumed to be a smooth function that decays rapidly at infinity. The material in this section is based on the work of Sloan and Osborn (1987). A similar method has been proposed by Sugihara (1987).

We remark that given an integral of the form

$$If = \int_{\Omega_s} g(\mathbf{t})\,d\mathbf{t},$$

where Ω_s is a bounded or unbounded domain in $\mathbb{R}^s$ with piecewise-smooth boundary, and g is a smooth function in Ω_s with possible integrable singularities on its boundary, it may be possible to use a nonlinear transformation to transform the integral into one of the form (9.16). For example, if Ω_s is an s-sphere of radius σ centred at the origin, then one may use a radial transformation proposed by Sag and Szekeres (1964),

$$\|\mathbf{t}\| = \sigma \tanh\left(u\|\mathbf{x}\|/\left(1 - \|\mathbf{x}\|^2\right)\right), \quad \|\mathbf{x}\| < 1,$$

where $u > 0$ is a parameter. This transformation maps Ω_s on to the unit sphere $\{\mathbf{x} : \|\mathbf{x}\| < 1\}$, and because its Jacobian and all its derivatives vanish as $\|\mathbf{x}\| \to 1$, the resulting integral with respect to $\mathbf{x}$ can be expressed in the form (9.16) by defining $f(\mathbf{x})$ to be zero for $\|\mathbf{x}\| \geqslant 1$.

This last example draws attention to the fact that the support of f in (9.16) need not be the whole of $\mathbb{R}^s$. In such a case f and as many as possible of its derivatives should vanish on the boundary of the support of f in order to make f as smooth a function over $\mathbb{R}^s$ as possible.

The integration rule that we shall investigate in the section is the equal-weight rule

$$Qf = \det L \sum_{\mathbf{x} \in L} f(\mathbf{x}), \tag{9.17}$$

where $L \subset \mathbb{R}^s$ is an infinite lattice as defined in Definition 2.1. We recall that $\det L$ is the volume of the unit cell (see Figure 2.3), or, equivalently, the reciprocal of the point density of the lattice. We shall call formula (9.17) a 'lattice rule'. Note that, unlike lattice rules for integration over the s-dimensional cube, the lattice rules introduced here do not require that L have $\mathbb{Z}^s$ as a subset; that is, we do not ask that L be an integration lattice (see Definition 2.2).

In practice, the infinite sum in (9.17) must be truncated, for example by ignoring all lattice points beyond some cut-off radius σ_0. Here we shall not consider the important practical question of how to choose σ_0.

In the work earlier in this book on lattice rules for the unit cube, we obtained an error expression as the sum of the Fourier coefficients of f belonging to the dual lattice. The next theorem gives a corresponding error expression for the lattice rule (9.17) when used to approximate the integral (9.16). Again we see the role of the dual lattice $L^\perp$ defined by Definition 2.6. We remark that because L need not be an integration lattice, the components of a point in the dual lattice need not be integers.

Theorem 9.9. *Let f be a continuous function integrable over $\mathbb{R}^s$ having the Fourier transform*

$$\hat{f}(\mathbf{w}) = \int_{\mathbb{R}^s} e^{-2\pi i \mathbf{w}\cdot\mathbf{x}} f(\mathbf{x})\, d\mathbf{x}, \quad \mathbf{w} \in \mathbb{R}^s.$$

Further, suppose that f satisfies

$$|f(\mathbf{x})| \leqslant c\,(1 + \|\mathbf{x}\|)^{-s-\delta}, \tag{9.18}$$

and

$$|\hat{f}(\mathbf{w})| \leqslant c_1\,(1 + \|\mathbf{w}\|)^{-s-\delta} \tag{9.19}$$

for some constants c, c_1, and $\delta > 0$, where $\|\cdot\|$ is the Euclidean norm. Then the error in the lattice rule (9.17) is given by

$$Qf - If = {\sum_{\mathbf{w}\in L^\perp}}' \hat{f}(\mathbf{w}), \tag{9.20}$$

where as usual, the prime on the sum indicates that the $\mathbf{w} = \mathbf{0}$ term is to be omitted.

Proof The proof is based on the Poisson summation formula

$$\sum_{\mathbf{m}\in\mathbb{Z}^s} f(\mathbf{m}) = \sum_{\mathbf{v}\in\mathbb{Z}^s} \hat{f}(\mathbf{v}),$$

proved by Stein and Weiss (1971, p.252) under assumptions (9.18) and (9.19). The sums on both sides of the equation converge absolutely.

For our proof we require a corresponding result for the lattice L, namely,

$$\det L \sum_{\mathbf{x}\in L} f(\mathbf{x}) = \sum_{\mathbf{w}\in L^\perp} \hat{f}(\mathbf{w}). \tag{9.21}$$

This may be obtained from the following argument.

Let A be a generator matrix for L. That is to say, L is the set of all integer linear combinations of the rows of A. Then we have seen in Section 2.11 that $(A^{-1})^T$ is a generator matrix for $L^\perp$, so L and $L^\perp$ are related to $\mathbb{Z}^s$ by the affine transformations

$$L = A^T \mathbb{Z}^s, \qquad L^\perp = A^{-1}\mathbb{Z}^s.$$

To prove (9.21), we write

$$\sum_{\mathbf{x}\in L} f(\mathbf{x}) = \sum_{\mathbf{m}\in\mathbb{Z}^s} f(A^T\mathbf{m}) = \sum_{\mathbf{m}\in\mathbb{Z}^s} F(\mathbf{m}),$$

where F is defined by $F(\mathbf{m}) := f(A^T\mathbf{m})$. From the definition of $\hat{F}$, it follows easily that

$$\hat{F}(\mathbf{v}) = |\det A|^{-1}\hat{f}\left(A^{-1}\mathbf{v}\right).$$

Moreover, F and $\hat{F}$ satisfy conditions (9.18) and (9.19) (possibly with different constants). Recalling that $\det L = |\det A|$, we obtain, using the Poisson summation formula,

$$\begin{aligned}
\det L \sum_{\mathbf{x}\in L} f(\mathbf{x}) &= \det L \sum_{\mathbf{m}\in\mathbb{Z}^s} F(\mathbf{m}) = \det L \sum_{\mathbf{v}\in\mathbb{Z}^s} \hat{F}(\mathbf{v}) \\
&= \sum_{\mathbf{v}\in\mathbb{Z}^s} \hat{f}\left(A^{-1}\mathbf{v}\right) = \sum_{\mathbf{w}\in L^\perp} \hat{f}(\mathbf{w}),
\end{aligned}$$

thus proving (9.21).

Since the left-hand side of (9.21) is Qf, and since the $\mathbf{w} = \mathbf{0}$ term of the right-hand side is $\hat{f}(\mathbf{0}) = If$, it follows immediately that

$$Qf - If = {\sum_{\mathbf{w}\in L^\perp}}' \hat{f}(\mathbf{w}),$$

completing the proof. ∎

In principle, the error expression (9.20) gives us a way of choosing a lattice to use in (9.17): clearly, we should choose L so that the nonzero points of $L^\perp$ occur only where $\hat{f}(\mathbf{w})$ is small. If $\hat{f}(\mathbf{w})$ decays rapidly as $\|\mathbf{w}\| \to \infty$, then one way of achieving this is to increase the scale of $L^\perp$, or, equivalently, to decrease the scale of L. This increases the computational work, dramatically so if s is large, so a significant change of scale will often not be possible when s is large. An alternative is to arrange the geometrical distribution of points of $L^\perp$ in a better way, with the value of $\det L$ held fixed. This is the next item for discussion.

Since the best choice of lattice will depend on the properties of the integrand f, it is necessary to make some assumptions about f. Our general assumptions are that f, and hence $\hat{f}$, exhibit essentially the same behaviour in each direction, and that $\hat{f}(\mathbf{w})$ decays in some suitable way at infinity. In this situation it is natural to express the error as a sum of contributions from successive spherical shells of $L^{\perp}$. If we let $\mu_1, \mu_2, \ldots$ be the radii of successive shells formed by the nonzero points of $L^{\perp}$, and let S_j be the shell of radius μ_j, $j = 1, 2, \ldots$, the error expression (9.20) may be written as

$$Qf - If = \sum_{j=1}^{\infty} \sum_{\mathbf{w} \in S_j} \hat{f}(\mathbf{w}). \tag{9.22}$$

The next result is an immediate consequence of Theorem 9.9. In it τ_j denotes the number of points of $L^{\perp}$ that lie on the shell S_j.

Corollary   Let f be a continuous function satisfying (9.18), whose Fourier transform satisfies

$$|\hat{f}(\mathbf{w})| \leqslant \phi\left(\|\mathbf{w}\|\right) \leqslant c_1 \left(1 + \|\mathbf{w}\|\right)^{-s-\delta}$$

for some real-valued function ϕ. Then the error in the lattice rule (9.17) satisfies

$$|Qf - If| \leqslant \sum_{j=1}^{\infty} \tau_j \phi(\mu_j) < \infty.$$

A case of particular interest is that in which $\hat{f}(\mathbf{w})$ decays at least exponentially as $\|\mathbf{w}\| \to \infty$. Suppose that $\hat{f}$ satisfies

$$|\hat{f}(\mathbf{w})| \leqslant Ce^{-\alpha\|\mathbf{w}\|}, \tag{9.23}$$

for some numbers C and α. It then follows from the corollary that

$$|Qf - If| \leqslant C \sum_{j=1}^{\infty} \tau_j e^{-\alpha\mu_j}. \tag{9.24}$$

Since the first term on the right-hand side is increasingly dominant as $\alpha \to \infty$, it follows that for α in (9.23) sufficiently large *the better of two lattices in the sense of giving a smaller bound in (9.24) is the one that has the larger value of* $\mu := \mu_1$. For two lattices with equal values of μ, the better lattice, in the same sense, is the one with the smaller value of $\tau := \tau_1$.

To formalize this result, it is convenient to define a class of functions $E_\alpha(C)$ as follows: a continuous function defined on $\mathbb{R}^s$ belongs to $E_\alpha(C)$ if it satisfies both (9.18) and (9.23). For functions in $E_\alpha(C)$ the bound

(9.24) holds. Thus if one has to choose between two lattices for a function in $E_\alpha(C)$, an appropriate measure of their quality is

$$R_\alpha = \sum_{j=1}^{\infty} \tau_j e^{-\alpha\mu_j},\qquad(9.25)$$

which is the error bound for the case $C = 1$. For any $\epsilon > 0$, it is clear that for α sufficiently large

$$R_\alpha \leqslant (1 + \epsilon)\tau e^{-\alpha\mu}.$$

It then follows that the better lattice for large α is the one with the larger value of μ.

A similar result holds if the exponent α in (9.23) is held constant, but the scale of the lattice L is made sufficiently small. To see this, it is convenient to introduce the lattice $\bar{L}$, which is defined as the lattice L scaled so that its determinant is 1. Then we may write $L = h\bar{L}$, where $h > 0$ is such that $h^s = \det L$. In turn we have $L^\perp = h^{-1}\bar{L}^\perp$, from which it follows that

$$\mu_j = h^{-1}\bar{\mu}_j, \quad j = 1, 2, 3\ldots,$$

where $\bar{\mu}_j$ is the radius of the jth shell of the dual lattice of $\bar{L}$. Then (9.24) may be written as

$$|Qf - If| \leqslant C\sum_{j=1}^{\infty} \tau_j e^{-(\alpha/h)\bar{\mu}_j}.$$

Clearly, the term of lowest order in h is the first. It is also clear that the larger the value of $\bar{\mu} = \bar{\mu}_1$, the faster the rate of approach of the first term to zero.

More generally, we may think of both α and h as varying. The above discussion is summarized in the following theorem.

Theorem 9.10. *Let $\bar{L}$ be a lattice with determinant 1. If $L = h\bar{L}$ and $f \in E_\alpha(C)$, then the error in the lattice rule (9.17) satisfies*

$$|Qf - If| \leqslant C\sum_{j=1}^{\infty} \tau_j e^{-(\alpha/h)\bar{\mu}_j}.$$

Given any $\epsilon > 0$, the error is bounded by

$$|Qf - If| \leqslant C(1 + \epsilon)\tau e^{-(\alpha/h)\bar{\mu}_j}$$

if α/h is sufficiently large. Hence given two lattices $\bar{L}_1$ and $\bar{L}_2$ with determinant 1, the better lattice, in the sense of the above bound and for α/h sufficiently large, is the one with the larger value of $\bar{\mu}$.

It turns out that we can express these conclusions in the language of sphere packings (for example, see Sloane 1981). Associated with $L^{\perp}$ is a packing of spheres of radius $\mu/2$, since spheres of this radius can obviously be centred at each point of $L^{\perp}$ without overlapping, but with each sphere being touched by some of its neighbours. The 'kissing number' of the sphere packing, the number of spheres that touch the sphere centred at the origin, is given by τ. The density of the sphere packing, which is roughly the fraction of the total volume that is occupied by the spheres, is given by

$$V_s(\mu/2)^s / \det L^{\perp} = \det L \times V_s(\mu/2)^s,$$

where

$$V_s = \frac{\pi^{s/2}}{\Gamma(\tfrac{1}{2}s + 1)}$$

is the volume of the unit sphere in s dimensions (so that $V_s(\mu/2)^s$ is the volume of the sphere of radius $\mu/2$).

In this terminology, Theorem 9.10 expresses the conclusion that for α sufficiently large, or h sufficiently small, *the better of two lattices L_1 and L_2 is the one whose dual has the denser sphere packing.* For $s = 2, 3, 4, 5,$ 8, the lattices with the densest possible sphere packings are known, while for $s = 6, 7, 12, 16, 24$, the lattices with densest known sphere packings are available. Further details may be found in Sloane (1981). In principle, the dual lattices of each of these lattices are good candidates for use in the lattice rule (9.17).

It now remains to consider functions whose Fourier transforms do not decay at the fast exponential rate already considered. In intermediate situations, there may be merit in evaluating sums such as (9.25) to allow a quantitative comparison. On the other hand, in the case of functions f so badly behaved that $\hat{f}(\mathbf{w})$ varies little over distances of order h^{-1}, it seems likely that any reasonable lattice with a given determinant will be as good as any other; for in this situation one would expect the error expression (9.20) to pick up many points of $L^{\perp}$ with equal weight. Thus the only aspect of $L^{\perp}$ that is likely to matter is the density of the points, rather than their distribution.

It would therefore seem to be sound policy in all situations to use lattices L with densely packed dual lattices $L^{\perp}$. In favourable cases, such a choice will yield positive benefits, while in unfavourable cases it should cause no disadvantage.

After having made a choice of lattice, one has to be able to generate the lattice points required in the lattice rule (9.17). This can be done by making use of the generator matrix of the lattice. For a given cut-off radius σ_0, it is possible to use the generator matrix to find the lattice points which are at a distance $\leqslant \sigma_0$ from the origin. An algorithm for this may be found in Joe (1989). The algorithm essentially involves solving sets of Diophantine equations by exhaustive search. The timing tests in this work indicate that generation of the lattice points should not usually be a major computational overhead in comparison to the function evaluations. Generator matrices for some interesting lattices with densely packed duals may be found in Sloane (1981).

10
Practical implementation of lattice rules

10.1 Introduction

In previous chapters we have been mainly concerned with the theoretical properties of lattice rules. Now in this chapter we turn to practical matters. We recall that Chapter 6 gave theoretical grounds for believing that it would be advantageous to use lattice rules of maximal rank. Chapter 7 showed that there might also be some merit in using certain lattice rules of intermediate rank, with the relative advantage increasing progressively with rank. To benefit from these results in a practical way, this chapter will look at a sequence of embedded lattice rules, with the successive rules having all ranks from 1 to s inclusive. Each rule contains all of the points of the preceding rule, and (in the preferred situation) an equal number of new ones. The final rule in this sequence, the one with the most lattice points, is a maximal rank rule. The details of the explicit construction and properties of such embedded sequences may be found in the next section.

Embedded sequences of quadrature rules open up the possibility of obtaining an error estimate with little extra cost. The hope is that such an error estimate, along with the full set of approximate integrals obtained from the sequence of embedded rules, can give valuable information about the reliability of the final (and hopefully best) approximation in the sequence. This topic is taken up in Section 10.3, and a practical scheme is put forward which makes use of the fact that the final lattice rule contains many different embedded rules with significant numbers of points.

An alternative error estimation technique discussed here is a generalization of the randomization procedure described earlier in Section 4.6 for number-theoretic rules. However, the cost of that technique is inevitably high, as the total number of function evaluations is increased by a significant factor.

Section 10.4 gives an algorithm allowing the calculation of all the approximations in the embedded sequence to be carried out simultaneously. This algorithm incorporates the error estimation procedure already described in Section 10.3.

The embedded sequences considered here require the choice of an s-dimensional integer vector $\mathbf{z}$. In Section 10.5 we shall discuss some of the practical details of the computer searches that are required to find such vectors. We have seen in Chapter 6 that each test of a particular $\mathbf{z}$ vector can be carried out with many fewer function evaluations than will be used in the final rule. Even so, as has already been indicated in Chapter 4, it is not usually feasible to search through all integer vectors. Thus it is common to limit the search to vectors $\mathbf{z}$ of a restricted form. One form that we have seen before (in Chapters 4 and 5) is the one-parameter form used by Korobov (1960). In Section 10.6 we give a result showing how searches using this form may be speeded up by omitting lattice rules that are geometrically equivalent. The proof of this result uses elementary number theory.

10.2 Embedded lattice rules

The embedded sequences of lattice rules that we shall consider are given by

$$
Q_r f = \frac{1}{n^r m} \sum_{k_r=0}^{n-1} \cdots \sum_{k_1=0}^{n-1} \sum_{j=0}^{m-1} f\left(\left\{ \frac{j}{m}\mathbf{z} + \frac{(k_1,\ldots,k_r,0,\ldots,0)}{n} \right\}\right), \quad (10.1)
$$

for $0 \leqslant r \leqslant s$, where n and m are relatively prime, and $\mathbf{z}$ is an integer vector having no nontrivial factor in common with m. It is clear that this sequence of lattice rules is an embedded sequence, since the points of Q_r are also points of Q_{r+1} for $0 \leqslant r \leqslant s - 1$. It follows from Theorem 7.1 that Q_r is a lattice rule having $n^r m$ points. Moreover, Q_0 is a rank-1 rule, while for $1 \leqslant r \leqslant s$, Q_r has rank r.

Example 10.1. *In Figures 10.1–10.3 we show the quadrature points for the two-dimensional embedded sequence with parameters*

$$
s = 2, \qquad n = 2, \qquad m = 5, \qquad \mathbf{z} = (1, 2).
$$

The points of Q_1 are those of Q_0 plus an equal number of new ones. Similarly, the points of Q_2 are those of Q_1 plus an equal number of new ones. Both Q_0 and Q_1 are of rank 1; they are generated by $(\frac{1}{5}, \frac{2}{5})$ and $(\frac{1}{10}, \frac{2}{10})$ respectively. The rule Q_2 is of rank-2, since it is the 2^2 copy of the rule in the small square: it is the 2^2 copy of Q_0! The reader might like to check that the behaviour is quite different if the value of m is changed to 4, which violates the condition that n and m be relatively prime.

We now show generally that, provided n and m are relatively prime, Q_s is the n^s copy of the rank-1 rule Q_0. Thus the embedded sequence given in (10.1) may be viewed as a particular approach to the evaluation of the n^s copy of Q_0.

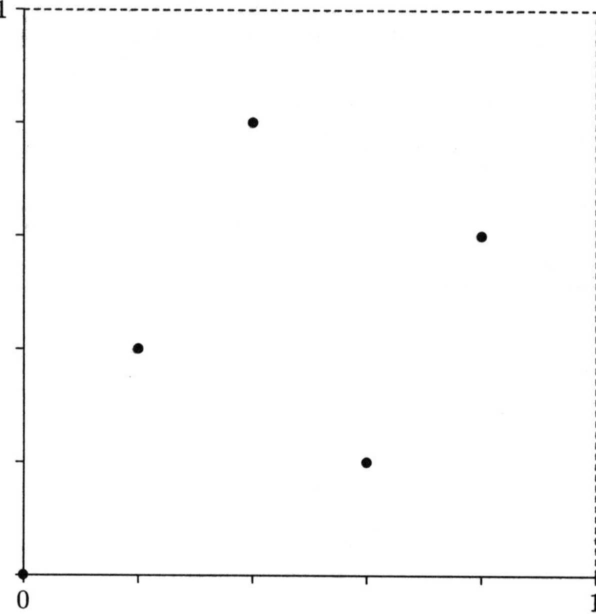

Fig. 10.1 Q_0.

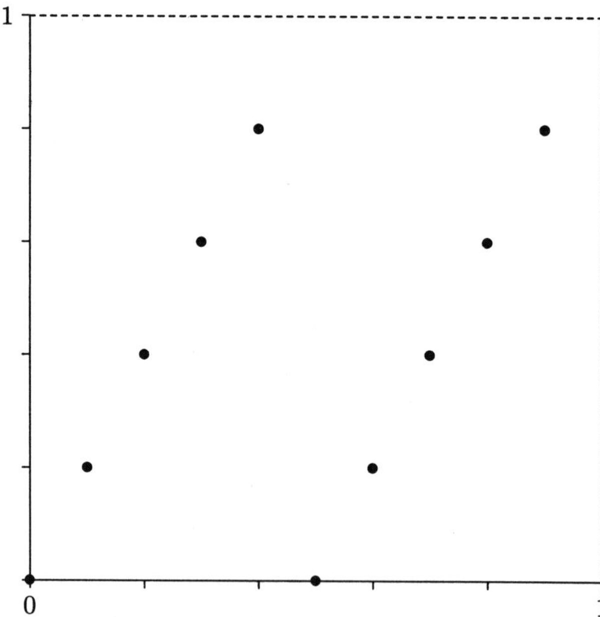

Fig. 10.2 Q_1.

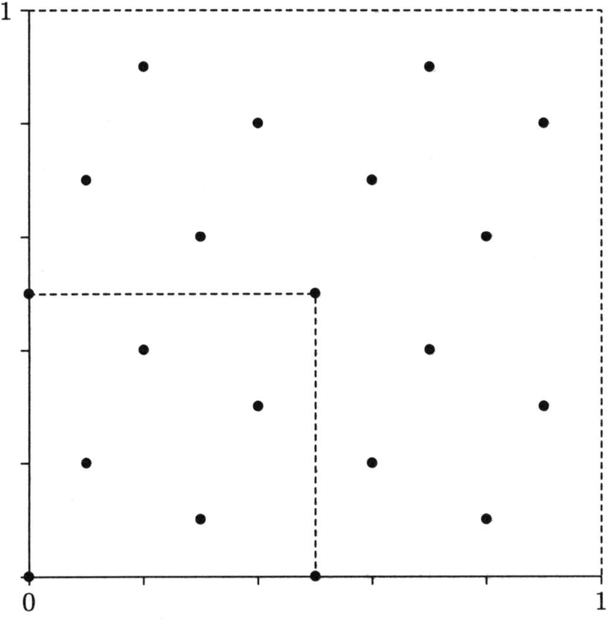

Fig. 10.3 Q_2.

Proposition 10.2. *If n and m are relatively prime, the lattice rule Q_s may be written as*

$$Q_s f = \frac{1}{n^s m} \sum_{k_s=0}^{n-1} \cdots \sum_{k_1=0}^{n-1} \sum_{j=0}^{m-1} f\left(\left\{\frac{j}{nm}\mathbf{z} + \frac{(k_1,\ldots,k_s)}{n}\right\}\right); \qquad (10.2)$$

that is, Q_s is the n^s copy of the rule Q_0.

Proof We need to show that (10.1) (with $r = s$) and (10.2) are equivalent. The right-hand side of (10.2) is, by definition, the n^s copy of the rule Q_0 applied to f. It is nonrepetitive by construction, and is a lattice rule of rank s by Theorem 6.4. Now the two lattice rules given by (10.1) (with $r = s$) and the right-hand side of (10.2) each have the same order $n^s m$, Moreover, the lattice corresponding to Q_s, namely,

$$L = \left\{\frac{j}{m}\mathbf{z} + \frac{(k_1,\ldots,k_s)}{n} : j, k_i \in \mathbb{Z}, \quad 1 \leqslant i \leqslant s\right\},$$

is clearly a subset of the lattice which corresponds to the right-hand side of (10.2), namely,

$$L' = \left\{\frac{j}{nm}\mathbf{z} + \frac{(k_1,\ldots,k_s)}{n} : j, k_i \in \mathbb{Z}, \quad 1 \leqslant i \leqslant s\right\}.$$

Since L is a subset of L' and each lattice has the same number of points per unit volume (namely $n^s m$), they must coincide, from which it follows that Q_s may be expressed equivalently in the form (10.2). ∎

As each successive lattice rule in the embedded sequence (10.1) has n times as many points as the preceding one, we might expect the lattice rules to get better as r increases. In terms of the quantity P_α defined in (5.1) and (5.2) this is certainly the case.

Proposition 10.3. *Let $\alpha > 1$. Then the lattice rules Q_r defined by (10.1) satisfy*

$$P_\alpha(Q_0) > P_\alpha(Q_1) > \cdots > P_\alpha(Q_s).$$

Proof The result follows from Theorem 5.1 and the obvious inclusions

$$L(Q_0) \subset L(Q_1) \subset \cdots \subset L(Q_s). \tag{10.3}$$

(The inequalities are strict as the inclusions in (10.3) are strict.) ∎

A key question to be decided before the embedded sequence can be used in practice is the choice of a value of n. A strong case can be made, on the grounds of both theory and practice, that the best value is $n = 2$. On the side of theory, Theorem 6.9 establishes that there exist very good rules that are 2^s copies of rank-1 rules. In the present context this means that if we set $n = 2$ then there exist very good candidates to serve as the rule Q_s that lies at the end of the sequence. On the side of practice, embedded sequences are very attractive when the order merely doubles at each step, instead of increasing by some higher multiple. With this choice there is always available a comparison rule with half the number of points, which is a large enough fraction to be useful. And in fact in the present context the choice $n = 2$ makes available to us (as we shall see in the next section) s different rules each having half the order of Q_s.

Once n has been determined, the question of choosing m (aside from the constraint that it be relatively prime with n—which means that it must be odd if $n = 2$) is essentially a question of deciding how many function evaluations one is prepared to allow. Finally, the choice of $\mathbf{z}$ should be such as to make Q_s as effective a rule as possible. We shall return with precise guidance on how to choose $\mathbf{z}$ in Section 10.5.

10.3 Error estimation

In one-dimensional numerical integration, a common way of obtaining an error estimate is to use two rules, say $\hat{Q}$ and $\tilde{Q}$, and to estimate the error by the difference $|\hat{Q}f - \tilde{Q}f|$, perhaps multiplied by some constant. (For a more sophisticated development of the same idea using the concept of 'null rules', see Berntsen and Espelid 1991.) Since the sequence in (10.1) is an embedded sequence and the members Q_{s-1} and Q_s use the most points, a natural first idea, proposed in Joe and Sloan (1992a), is to use as an error estimate the quantity

$$c|Q_s f - Q_{s-1} f|, \tag{10.4}$$

where c is some scaling constant. As will be seen in the next section, all the members Q_r, $0 \leqslant r \leqslant s$, of the embedded sequence may be calculated simultaneously, so that calculation of the error estimate (10.4) would require negligible extra computation. However, numerical tests based on the testing package of Genz (1984, 1987) indicate that the error estimate (10.4) (with $c = 2$), though reasonably reliable in low dimensions, tends to underestimate the true error as the dimension s gets larger. The problem, of course, is that it can happen only too easily that $Q_{s-1} f$ and $Q_s f$ have similar values, even when neither result is particularly close to the true value. And we certainly do not have any theorem to tell us that the convergence of the sequence $Q_0 f, \ldots, Q_s f$ to If is monotone, or regular in any way.

The improved error estimate proposed here is based on the recognition that the $n^s m$-point rule Q_s contains not just one embedded rule with $n^{s-1} m$ points, but s such rules, which differ from Q_{s-1} only in the choice of the coordinate that 'misses out' on being summed over. In (10.1) with $r = s - 1$, the distinguished coordinate is the sth, but it may as well be any other. We would expect a more robust error estimate to be obtained by taking the root mean square of the error estimates from each such choice. This estimate is like (10.1) in lacking a theoretical basis, but it is undeniably safer, with the safety net improving as the dimension s increases.

In more detail, we first define for $1 \leqslant i \leqslant s$ the lattice rules

$$Q^{(i)} f := \frac{1}{n^{s-1} m} \sum_{k_s=0}^{n-1} \cdots \sum_{k_{i+1}=0}^{n-1} \sum_{k_{i-1}=0}^{n-1} \cdots \sum_{k_1=0}^{n-1} \sum_{j=0}^{m-1} f\left(\left\{ \frac{j}{m} \mathbf{z} + \frac{\mathbf{k}^{(i)}}{n} \right\} \right),$$

where $\mathbf{k}^{(i)} = (k_1, \ldots, k_{i-1}, 0, k_{i+1}, \ldots, k_s)$. Thus $Q^{(i)}$ is a lattice rule of order $n^{s-1} m$, and $Q^{(s)} = Q_{s-1}$. It is apparent that the points of $Q^{(i)}$ are also points of Q_s, so that $Q^{(i)}$ is embedded in Q_s for $1 \leqslant i \leqslant s$.

Since $Q_s f$ is our final approximation, we can obtain s error estimates by calculating

$$Q_s f - Q^{(i)} f, \quad 1 \leqslant i \leqslant s.$$

To combine these s error estimates into one, following Joe and Sloan (1993) we propose the error estimate given by

$$E = c \left(\sum_{i=1}^{s} \left(Q_s f - Q^{(i)} f \right)^2 / s \right)^{1/2}, \tag{10.5}$$

with c a 'fudge factor' allowing a safety margin. In the calculations in the next chapter, we take $c = 1$. Since the approximations $Q^{(1)} f, \ldots, Q^{(s)} f$

can be calculated at the same time as we calculate the embedded sequence $Q_0 f, \ldots, Q_s f$, the extra computation time required to obtain E is negligible. (Note that the s differences $Q_s - Q^{(i)}$ are 'null rules' in the sense of Berntsen and Espelid 1991, in that they all yield zero when applied to the constant function.)

The error estimate E just considered is relevant only for our special class of lattice rules. For general lattice rules an alternative error estimation strategy is to use a randomization procedure. This procedure is an extension of the one described in Section 4.6 for number-theoretic rules. Since the generalization is quite obvious, we shall be content to give only an outline.

Given the lattice rule

$$Qf = \frac{1}{N} \sum_{j=0}^{N-1} f(\mathbf{x}_j),$$

in which the points $\mathbf{x}_0, \ldots, \mathbf{x}_{N-1}$ are all the points of an integration lattice L that lie in the unit cube U^s, in the randomization procedure one uses the shifted lattice rule

$$Q_{\mathbf{c}} f = \frac{1}{N} \sum_{j=0}^{N-1} f\left(\{\mathbf{x}_j + \mathbf{c}\}\right)$$

with $\mathbf{c}$ a random vector. We see from (2.22) that the error for this shifted lattice rule is

$$Q_{\mathbf{c}} f - If = {\sum_{\mathbf{h} \in L^{\perp}}}' e^{2\pi i \mathbf{h} \cdot \mathbf{c}} \hat{f}(\mathbf{h}).$$

Analogous to Theorem 4.11 and its corollary, we then have the following.

Theorem 10.4. *Suppose $\mathbf{c}$ has a multivariate uniform distribution on C^s. Then*

$$E(Q_{\mathbf{c}} f) = If,$$

where $E(\cdot)$ is the expectation operator.

Corollary For any positive integer q, let $\mathbf{c}_1, \ldots, \mathbf{c}_q$ be independent random vectors with a multivariate uniform distribution on C^s. Then the estimate

$$\bar{Q} f = \frac{1}{q} \sum_{p=1}^{q} Q_{\mathbf{c}_p} f$$

is an unbiased estimate of If.

Hence we can obtain an unbiased approximation to If by taking the mean of q shifted lattice rules. As in Section 4.6, we can be $(100 - 100/v^2)\%$

'confident' that $v\sigma$ is an upper bound on the true error, where in practice σ is approximated by

$$\tilde{\sigma} = \sqrt{\frac{\sum_{p=1}^{q} \left(Q_{c_p} f - \bar{Q} f\right)^2}{q(q-1)}}.$$

Further details about this randomization procedure may be found in Joe (1990a). The problem with this procedure is its high cost, in that it rests on using the basic lattice rule q times.

10.4 Calculation of the embedded sequence simultaneously

Here we give an algorithm which implements the embedded sequence (10.1) and the error estimate (10.5) (with $c = 1$). This algorithm is the same as one appearing in Joe and Sloan (1993) except that a small error (affecting only the calculation of $Q_0 f$) has been corrected. The nested loops which would normally arise from the multiple sums in (10.1) have been unravelled in a manner similar to that used in the function SUM found in Davis and Rabinowitz (1984, pp.488–489). The resulting looping procedure is analogous to the operation of an odometer.

The variable ω in the algorithm labels the outermost sum in (10.1) for which, up to the present stage, any summation index has been changed from zero. The remaining summation indices are zero, that is, at every stage

$$k_\omega \neq 0, \qquad k_{\omega+1} = \cdots = k_s = 0.$$

Thus we know that the current function value contributes to $Q_\omega f, \ldots, Q_s f$, but (because $k_\omega \neq 0$) not to $Q_0 f, \ldots, Q_{\omega-1} f$.

comment: Variables used are:

$k_i, \ 1 \leqslant i \leqslant s$: Summation variables in (10.1).

$x_i, \ 1 \leqslant i \leqslant s$: $\mathbf{x} = (x_1, \ldots, x_s)$.

$z_i, \ 1 \leqslant i \leqslant s$: The vector $\mathbf{z} = (z_1, \ldots, z_s)$ in (10.1).

n : The parameter of this name in (10.1).

$Q_i, \ 0 \leqslant i \leqslant s$: Eventually contains $Q_i f$.

val : $\displaystyle\sum_{j=0}^{m-1} f\left(\left\{\frac{j}{m}\mathbf{z} + \frac{(k_1, k_2, \ldots, k_s)}{n}\right\}\right).$

i, j, ℓ : Loop counters.

ω : Described above, $0 \leqslant \omega \leqslant s$.

$Q^{(i)}, \ 1 \leqslant i \leqslant s$: Eventually contains $Q^{(i)} f$.

E : Eventually contains the error estimate given by (10.5).

Firstly, we initialize some variables.

endcomment

$$Q_0 = 0.0$$
for $i = 1$ **to** s **step** 1 **do:**
   $k_i = 0$
   $Q_i = 0.0$
   $Q^{(i)} = 0.0$
endfor
$$k_1 = -1$$
$$\ell = 1$$
$$\omega = 0$$
comment:
   Main body of algorithm follows.
endcomment
while $\ell \leqslant s$ **do:**
   if $k_\ell < n - 1$ **then do:**
     $k_\ell = k_\ell + 1$
     $\ell = 1$
     $val = 0.0$
     for $j = 0$ **to** $m - 1$ **step** 1 **do:**
       for $i = 1$ **to** s **step** 1 **do:**
$$x_i = \left(\frac{j}{m} z_i + \frac{k_i}{n} \right) \bmod 1$$
       endfor
      $val = val + f(\mathbf{x})$
     endfor
     for $i = \omega$ **to** s **step** 1 **do:**
      $Q_i = Q_i + val$
     endfor
     for $i = 1$ **to** s **step** 1 **do:**
      if $k_i = 0$ **then do:**
       $Q^{(i)} = Q^{(i)} + val$
      endif
     endfor
     if $\omega = 0$ **then do:**
      $\omega = 1$
     endif
   else do:
     $k_\ell = 0$
     $\ell = \ell + 1$
     if $\omega < \ell$ **then do:**
      $\omega = \ell$
     endif
   endif
endwhile

for $i = 0$ **to** s **step** 1 **do:**
  $Q_i = Q_i/(n^i m)$
endfor
$E = 0.0$
for $i = 1$ **to** s **step** 1 **do:**
  $Q^{(i)} = Q^{(i)}/(n^{s-1} m)$
  $E = E + (Q_s - Q^{(i)})^2$
endfor
$E = \sqrt{E/s}$

10.5   Searches for a good embedded sequence

Once we have chosen n (in practice usually 2 as indicated earlier) and m (which must be relatively prime to n, and so must be odd if $n = 2$), all that remains is to make a choice of the integer vector $\mathbf{z}$.

Now recall that our criterion for measuring the 'goodness' of a lattice rule is the quantity P_α, $\alpha > 1$. If the lattice rule Q has the corresponding lattice L, then we recall from (5.1) that

$$P_\alpha(Q) = \sum_{\mathbf{h} \in L^\perp}' \frac{1}{(\overline{h}_1 \overline{h}_2 \cdots \overline{h}_s)^\alpha};$$

it is an upper bound on the quadrature error for functions whose Fourier coefficients $\hat{f}$ satisfy

$$|\hat{f}(\mathbf{h})| \leqslant \frac{1}{(\overline{h}_1 \overline{h}_2 \cdots \overline{h}_s)^\alpha}, \qquad \mathbf{h} \in \mathbb{Z}^s.$$

In practice, $P_\alpha(Q)$ is calculated by using (see (5.2))

$$P_\alpha(Q) = Q \prod_{k=1}^{s} F_\alpha(x_k) - 1,$$

where

$$F_\alpha(x) = 1 + \sum_{h \neq 0} \frac{e^{2\pi i h x}}{|h|^\alpha}.$$

For convenience, we give from (4.14) to (4.16) the expressions for $F_2(x)$, $F_4(x)$, and $F_6(x)$ when $x \in [0, 1]$:

$$
\begin{aligned}
F_2(x) &= 1 + 2\pi^2(x^2 - x + 1/6), \\
F_4(x) &= 1 + \frac{\pi^4}{45}\left(1 - 30x^2(1 - x)^2\right), \\
F_6(x) &= 1 + \frac{2\pi^6}{945}(1 - 21x^2 + 105x^4 - 126x^5 + 42x^6).
\end{aligned}
$$

Since our approximation to If will be given by $Q_s f$, we naturally want Q_s to be as accurate as possible. We saw from the results of Disney and

Sloan (1992) in Chapter 6 that there exists a $\mathbf{z}$ which is good in the sense that it produces a small value of $P_\alpha(Q_s)$ when used in a 2^s copy rule. However, the proof was nonconstructive. Thus in practice computer searches are required to find a suitable $\mathbf{z}$.

A great advantage of using $P_\alpha(Q_s)$ as the criterion in a computer search is that although the rule Q_s is of order $n^s m$, Lemma 6.5 shows that

$$P_\alpha(Q_s) = Q_0 f_\alpha^{(n)} - 1,$$

where

$$f_\alpha^{(n)}(\mathbf{x}) = \prod_{k=1}^{s} F_\alpha^{(n)}(x_k), \tag{10.6}$$

with

$$F_\alpha^{(n)}(x) = 1 + \frac{1}{n^\alpha} \sum_{h \neq 0} \frac{e^{2\pi i h x}}{|h|^\alpha} = 1 + \frac{1}{n^\alpha}(F_\alpha(x) - 1).$$

Thus $P_\alpha(Q_s)$ may be calculated by using Q_0, a rule with only m lattice points.

We remark that since

$$F_\alpha^{(n)}(x) = F_\alpha^{(n)}(1 - x), \tag{10.7}$$

the number of terms in the sum for $P_\alpha(Q_s)$ can approximately be halved by writing

$$\frac{1}{m} \sum_{j=0}^{m-1} f_\alpha^{(n)}\left(\left\{\frac{j}{m}\mathbf{z}\right\}\right) = \frac{f_\alpha^{(n)}(\mathbf{0})}{m} + \frac{2}{m} \sum_{j=1}^{\lfloor m/2 \rfloor *} f_\alpha^{(n)}\left(\left\{\frac{j}{m}\mathbf{z}\right\}\right),$$

where the asterisk indicates that the term with $j = \lfloor m/2 \rfloor$ is to be halved in the event that m is even.

Practical considerations lead us to limit practical computer searches to vectors $\mathbf{z}$ of a restricted form. Here we recommend the Korobov form given in (4.32), namely,

$$\mathbf{z}(\ell) = (1, \ell, \ell^2 \bmod m, \ldots, \ell^{s-1} \bmod m), \qquad 1 \leqslant \ell \leqslant \lfloor m/2 \rfloor. \tag{10.8}$$

Thus there are now only $\lfloor m/2 \rfloor$ choices of $\mathbf{z}$ to search through.

Once we have an integer vector $\mathbf{z}$ that minimizes $P_\alpha(Q_s)$, we may wish to permute the components of $\mathbf{z}$ in the manner suggested by Joe and Sloan (1992a). Such a permutation will not affect $P_\alpha(Q_s)$ or the error estimate E given in (10.5). However, a permutation might be desirable in the hope of making the later estimates in the sequence $Q_0 f, \ldots, Q_{s-1} f$ (which use the most function evaluations) as accurate as possible.

The permutation procedure is as follows. Suppose we have done a computer search and found an integer vector $\mathbf{z}$ that minimizes $P_\alpha(Q_s)$. We now note that some permutations of the components of $\mathbf{z}$ affect $P_\alpha(Q_r)$: from (7.6) to (7.9) and (10.1) we see that $P_\alpha(Q_r)$ for $1 \leqslant r \leqslant s-1$ is unaffected by a permutation that affects only the first r components or the last $s-r$ components, but generally is affected by permutations that mix these two sets.

For this reason, instead of using $\mathbf{z}$, we may choose to use

$$\mathbf{z}' = (z_{i_1}, \ldots, z_{i_s}),$$

where $(i_1, \ldots, i_s)$ is a permutation of $(1, \ldots, s)$ determined as follows: first, choose i_s out of the s possible choices so that $P_\alpha(Q_{s-1})$ is minimized; then choose i_{s-1} out of the remaining $s-1$ choices so that $P_\alpha(Q_{s-2})$ is minimized; and so on until we come to i_2 which is chosen from the two remaining possibilities by minimizing $P_\alpha(Q_1)$. This permutation procedure requires a total of $(s(s+1)-2)/2$ calculations of $P_\alpha(Q_r)$ for $0 < r < s$ and a given $\mathbf{z}$. This is often a small number compared to the number of calculations of $P_\alpha(Q_s)$ required to find the original vector $\mathbf{z}$. We would, of course, use the result of Theorem 7.3, which allows $P_\alpha(Q_r)$, $1 \leqslant r \leqslant s-1$, to be calculated using a lattice rule with only m points instead of the usual $n^r m$ points. In this case the computation time required for the calculation of each $P_\alpha(Q_r)$ is essentially independent of r.

Tables of such $\mathbf{z}$ for $s = 2, \ldots, 12$ obtained by using the search procedure of this section together with the permutation option may be found in Appendix A.

10.6   Reducing the number of vectors z searched

We recall from the last section that if we were to restrict our computer searches to vectors $\mathbf{z}$ of the Korobov form (10.8), then there would be $\lfloor m/2 \rfloor$ such vectors to search through. In this section, we show that it is possible to further reduce the number of vectors to be searched, with the number being approximately halved in the best possible case (that is, when m is prime). As has already been mentioned, $P_\alpha(Q_s)$ is just the quadrature error obtained by applying the rank-1 rule Q_0 to the function $f_\alpha^{(n)}$. Now it is clear from (10.6) and (10.7) that $f_\alpha^{(n)}$ is unaffected by a permutation of its variables, or by a reflection in one or more mid-planes, or a combination of these. Thus two geometrically equivalent rank-1 rules would yield the same value of $P_\alpha(Q_s)$.

This leads to the main result of this section, which states that if ℓ_1 and ℓ_2 are integers satisfying $\ell_1\ell_2 \equiv \pm 1 \,(\mathrm{mod}\ m)$, then the rank-1 rules generated from the Korobov vectors $\mathbf{z}(\ell_1)$ and $\mathbf{z}(\ell_2)$ are geometrically equivalent.

Theorem 10.5. *Let* ℓ_1 *and* ℓ_2, $1 \leqslant \ell_1, \ell_2 \leqslant \lfloor m/2 \rfloor$, $m \geqslant 2$, *be integers satisfying*

$$\ell_1\ell_2 \equiv \pm 1 \,(\mathrm{mod}\, m), \tag{10.9}$$

and let f be a function defined on $\mathbb{R}^s$ which is one-periodic in each of its s variables. Then the rank-1 lattice rules

$$\frac{1}{m} \sum_{j=0}^{m-1} f\left(\left\{\frac{j}{m}\mathbf{z}(\ell_1)\right\}\right) \tag{10.10}$$

and

$$\frac{1}{m} \sum_{j=0}^{m-1} f\left(\left\{\frac{j}{m}\mathbf{z}(\ell_2)\right\}\right) \tag{10.11}$$

are geometrically equivalent.

Before giving the proof of Theorem 10.5, we need some elementary results from number theory. The simple proofs are included for completeness. For any two integers u and v, we denote the greatest common divisor of u and v by $\gcd(u, v)$.

Lemma 10.6. *Suppose $\ell_1\ell_2 \equiv \pm 1 \,(\mathrm{mod}\,m)$ with $m \geqslant 2$. Then*

(i) $\gcd(\ell_1, m) = 1$, *that is, ℓ_1 and m are relatively prime.*
(ii) $\gcd(\ell_1^k, m) = 1$ *for any integer $k \geqslant 1$.*

Proof The assumption in the lemma implies that there is an integer k such that $\ell_1\ell_2 - km = \pm 1$. Since any common divisor of ℓ_1 and m must therefore divide ± 1, we see immediately that the greatest common divisor of ℓ_1 and m must be 1. Part (ii) is a consequence of the first result, since any prime number that divides ℓ_1^k and m must also divide ℓ_1 (by the unique factorization theorem), and so must divide $\gcd(\ell_1, m) = 1$.   ∎

Lemma 10.7. *Let ℓ_1 and ℓ_2, $1 \leqslant \ell_1, \ell_2 \leqslant \lfloor m/2 \rfloor$, $m \geqslant 2$, be integers satisfying $\ell_1\ell_2 \equiv \pm 1 \,(\mathrm{mod}\,m)$. Also, let $\mathbb{Z}_m := \{j \in \mathbb{Z} : 0 \leqslant j \leqslant m - 1\}$. Then for $j \in \mathbb{Z}_m$, the mapping $\eta : \mathbb{Z}_m \to \mathbb{Z}_m$, given by*

$$\eta(j) = j\ell_1^{s-1} \bmod m,$$

is a one-to-one mapping.

Proof Suppose the contrary. Then there exist $j_1, j_2 \in \mathbb{Z}_m$, $j_1 \neq j_2$, such that $\eta(j_1) = \eta(j_2)$, that is, $j_1\ell_1^{s-1} \equiv j_2\ell_1^{s-1} \,(\mathrm{mod}\,m)$. It then follows that

$$(j_1 - j_2)\,\ell_1^{s-1} \equiv 0 \,(\mathrm{mod}\,m). \tag{10.12}$$

Since $1 \leqslant |j_1 - j_2| \leqslant m - 1$, and $\ell_1^{s-1} \neq 0$, it follows that for (10.12) to be satisfied, we must have $\gcd(\ell_1^{s-1}, m) > 1$, contradicting part (ii) of Lemma 10.6. Thus the result holds by contradiction.   ∎

Armed with these results, we can now give the proof of Theorem 10.5.

Proof of Theorem 10.5 Recall that $\mathbb{Z}_m = \{j \in \mathbb{Z} : 0 \leqslant j \leqslant m - 1\}$. Then for $j \in \mathbb{Z}_m$, the jth quadrature point of the quadrature rule (10.10) is given by $\{j\mathbf{z}(\ell_1)/m\}$. This point has kth component $\{j\ell_1^{k-1}/m\}$ for $1 \leqslant k \leqslant s$, where $\{x\}$ denotes the fractional part of x. Now the reflection of this component in the mid-plane perpendicular to the kth axis is given by

$$1 - \{j\ell_1^{k-1}/m\} = \{-j\ell_1^{k-1}/m\} \text{ if } j \neq 0,$$

while if $j = 0$ the component is left unchanged. Thus one way to prove that (10.10) and (10.11) are geometrically equivalent rules is to prove that for each $j \in \mathbb{Z}_m$ there exists a one-to-one mapping $\eta : \mathbb{Z}_m \to \mathbb{Z}_m$ such that, for $1 \leqslant k \leqslant s$, there is a component of $\{\eta(j)\mathbf{z}(\ell_2)/m\}$ which is equal to either $\{j\ell_1^{k-1}/m\}$ or $\{-j\ell_1^{k-1}/m\}$.

We now show that this can be done by taking

$$\eta(j) = j\ell_1^{s-1} \bmod m,$$

which, we recall from Lemma 10.7, is a one-to-one mapping. In particular, we shall prove that the $(s - k + 1)$th component of $\{\eta(j)\mathbf{z}(\ell_2)/m\}$ satisfies the required condition, that is, we shall prove that

$$\text{either } \{(j\ell_1^{s-1} \bmod m)\, \ell_2^{s-k}/m\} \;=\; \{j\ell_1^{k-1}/m\} \qquad (10.13)$$
$$\text{or } \{(j\ell_1^{s-1} \bmod m)\, \ell_2^{s-k}/m\} \;=\; \{-j\ell_1^{k-1}/m\}. \qquad (10.14)$$

Because $\ell_1 \ell_2 \equiv \pm 1 \,(\mathrm{mod}\, m)$, it follows that $\ell_2 = (um \pm 1)/\ell_1$ for some integer u. We then have, for any $j \in \mathbb{Z}_m$,

$$
\begin{aligned}
\{(j\ell_1^{s-1} \bmod m)\, \ell_2^{s-k}/m\} &= \{j\ell_1^{s-1}\ell_2^{s-k}/m\} \\
&= \left\{ j\ell_1^{s-1} \frac{(um \pm 1)^{s-k}}{m\ell_1^{s-k}} \right\} \\
&= \{j\ell_1^{k-1}(um \pm 1)^{s-k}/m\}.
\end{aligned}
$$

Now it follows from the binomial theorem that $(um \pm 1)^{s-k} = vm + (\pm 1)^{s-k}$ for some integer v. Thus

$$
\begin{aligned}
\{(j\ell_1^{s-1} \bmod m)\, \ell_2^{s-k}/m\} &= \{j\ell_1^{k-1}\left(vm + (\pm 1)^{s-k}\right)/m\} \\
&= \{(\pm 1)^{s-k} j\ell_1^{k-1}/m\},
\end{aligned}
$$

which yields either (10.13) or (10.14) depending on the parity of $s - k$. This completes the proof. $\blacksquare$

In a search through vectors of the Korobov form for the one that minimizes $P_\alpha(Q_s)$, Theorem 10.5 shows that we do not normally have to search through all the possible $\lfloor m/2 \rfloor$ choices. It follows from Niven and Zuckerman (1980, Theorem 2.13) that the equation $\ell v \equiv 1 \,(\mathrm{mod}\, m)$ has a solution if and only if $\gcd(\ell, m) = 1$. Moreover, if this condition is satisfied then there exists one and only one solution v^* belonging to $\mathbb{Z}_m$. Further, under the same condition $\gcd(\ell, m) = 1$ the equation $\ell v \equiv -1 \,(\mathrm{mod}\,m)$ also has a unique solution, modulo m, given by $m - v^*$. Since exactly one of v^* and $m - v^*$ is in the interval from 1 to $\lfloor m/2 \rfloor$ (unless they are both equal to $\lfloor m/2 \rfloor$), we conclude that if ℓ_1, $1 \leqslant \ell_1 \leqslant \lfloor m/2 \rfloor$, is relatively prime to m, then there exists a unique ℓ_2, $1 \leqslant \ell_2 \leqslant \lfloor m/2 \rfloor$, such that (10.9) holds, with one or other sign.

We are interested in the cases when $\ell_1 \neq \ell_2$. So for a given value of m, let μ_m be the number of such pairs ℓ_1 and ℓ_2, $\ell_1 \neq \ell_2$, such that $\ell_1 \ell_2 \equiv \pm 1 \,(\mathrm{mod}\,m)$ and $1 \leqslant \ell_1, \ell_2 \leqslant \lfloor m/2 \rfloor$. Then there are only $\hat{\mu}_m = \lfloor m/2 \rfloor - \mu_m$ choices of $\mathbf{z}(\ell)$ to search through rather than the original $\lfloor m/2 \rfloor$. In general, for given m, the value of μ_m is not known *a priori*. However, if m is a prime number, then $\gcd(\ell, m) = 1$ for all ℓ satisfying $1 \leqslant \ell \leqslant \lfloor m/2 \rfloor$. Hence, in this case (10.9) always has a solution ℓ_2 satisfying $1 \leqslant \ell_2 \leqslant \lfloor m/2 \rfloor$, and since $\ell_2 = \ell_1$ occurs not more than twice (see Niven and Zuckerman 1980, Corollary 2.28), the number of possible choices of $\mathbf{z}(\ell)$ is approximately halved.

We now describe a simple algorithm which allows us to make use of Theorem 10.5 and the ensuing discussion. The idea is that before the actual computer search is started, we find and store (in an array) the $\hat{\mu}_m$ values of ℓ for which a calculation of $P_\alpha(Q_s)$ using $\mathbf{z}(\ell)$ is required. Let this array be A, with array elements A_j. To compute the array, set $A_1 = 1$, $A_2 = 2$, and initialize the integer j to be 3. Suppose at the current value of j, the array A already contains t_{j-1} elements (with $t_2 = 2$). Then for $2 \leqslant k \leqslant t_{j-1}$, we test whether $A_k j \equiv \pm 1 \,(\mathrm{mod}\,m)$. If we find any such k, we set $t_j = t_{j-1}$. Otherwise, we set $t_j = t_{j-1} + 1$ and $A_{t_j} = j$. The integer j is then incremented by 1. This process is then continued until j reaches $\lfloor m/2 \rfloor + 1$, at which stage we have $t_{\lfloor m/2 \rfloor} = \hat{\mu}$.

Timing tests for this simple-minded algorithm given in Joe (1990b) indicate that it is quite effective. Suggestions for some other algorithms that are more efficient (but also more complicated) may also be found in Joe (1990b).

11

Comparisons with other methods

11.1 Introduction

In Chapter 6 we saw theoretical evidence (based on the criterion P_α) that maximal rank lattice rules obtained by copying rank-1 lattice rules are at least comparable in accuracy with rank-1 rules of similar order. Also, we saw that such maximal rank lattice rules have two practical advantages. Firstly, computer searches to find good rules are computationally less demanding than searches for good rank-1 rules, with this advantage increasing as the dimension s increases and being very substantial when s is large. Secondly, the previous chapter showed how we could use the lattice rules of rank $s - 1$ embedded in the maximal rank rule to obtain an estimate of the error. This error estimate requires very little extra computation. However, these advantages would be of little use in practical calculations unless the accuracy of such rules is comparable with that of rank-1 rules in practice, and unless the error estimate is reasonably reliable. And, lattice rules of any stamp would not be very useful in practice if they compared poorly with existing methods such as the Monte Carlo method.

To enable a comparison to be made between different methods, in this chapter we give extensive numerical tests obtained using our implementation of the testing package of Genz (1984, 1987). This testing package is described in more detail in the next section. With this package, we tested the following methods:

(C) An embedded sequence of lattice rules as implemented in the algorithm given in Section 10.4. The final rule in the sequence is the n^s copy of a rank-1 rule with the value of n taken here to be $n = 2$. The vectors $\mathbf{z}$ that are required in (10.1) were obtained by finding the vector of the Korobov form (10.8) that minimized P_2 for the final rule in the sequence, using the trick described in Section 10.5. The error estimate is that given in (10.5) with $c = 1$.

(M) An adaptive Monte Carlo method. Here we used the implementation found in the NAG routine D01GBF (Numerical Algorithms Group 1991). This being an adaptive scheme, one would expect it to be better than the straight Monte Carlo method.

(A) An adaptive method based on the algorithm of Van Dooren and De Ridder (1976) in conjunction with the modification of Genz and Malik (1980). Here we used the implementation found in the NAG routine D01FCF (Numerical Algorithms Group 1991).

(R) Rank-1 rules. Where available, the vectors $\mathbf{z}$ used to generate the rank-1 rules were the ones built into the NAG routine D01GCF (Numerical Algorithms Group 1991). These were obtained by a search of vectors of the Korobov form (10.8) to find the one that minimized the quantity $\tilde{P}_2(\mathbf{z}, N)$ given by (4.17). This search procedure is implemented in the NAG routine D01GYF (Numerical Algorithms Group 1991). However, the $\mathbf{z}$ built into D01GCF are available only for $N = 5003$, $10\,007$, $20\,011$, $40\,009$ and $80\,021$ (and also for $N = 2129$, though we did not use this value of N in our numerical tests). For the numerical tests with $N = 160\,001$, the NAG routine D01GYF was used to obtain the required $\mathbf{z}$. For values of N larger than this, the use of D01GYF was not feasible, because of the computing time required. Instead, an alternative routine D01GZF was used. This routine is much faster than D01GYF, and is suitable for use when N is a product of two distinct primes. The price is that the $\mathbf{z}$ obtained from this routine tend not to be as good as those from D01GYF (as measured by $\tilde{P}_2(\mathbf{z}, N)$). The routine D01GZF was used to obtain $\mathbf{z}$ for $N = 320\,423 = 4\,513 \times 71$, $N = 641\,423 = 7\,207 \times 89$, $N = 1\,249\,903 = 11\,467 \times 109$, and $N = 2\,500\,387 = 18\,251 \times 137$. In our rank-1 calculations, the randomization procedure (as described in Section 4.6) was not implemented, and hence no error estimate was generated from the rank-1 calculations. In effect, we are allowing the rank-1 rules every opportunity to shine in the competition for accuracy achieved with a given number of function evaluations, by this suppression of the error estimate. (We may recall that the process of error estimation by randomization increases the number of function evaluations by a factor of three to five.)

For ease of identifying the four methods, the methods C, M, A, and R will be called COPY, MCARLO, ADAPT, and RANK1 respectively.

A detailed description of the testing procedure used may be found in the next section. The numerical results obtained may be found in Section 11.3, while some conclusions may be found in the final section.

11.2   Testing procedure

We now describe the testing procedure used to obtain the numerical results given in the next section. We begin by looking at the testing package of Genz (1984, 1987). Full details of this package may be found in these two references, so we shall be content with an outline of the main points. The package is based on a collection of six families of test integrands. Each of these families is given a name or 'attribute'; along with their attributes, the six families are

1. OSCILLATORY: $\quad f^{(1)}(\mathbf{x}) \;=\; \cos\!\left(2\pi w_1 + \sum_{i=1}^{s} c_i x_i\right),$

2. PRODUCT PEAK: $\quad f^{(2)}(\mathbf{x}) \;=\; \prod_{i=1}^{s}\left(c_i^{-2} + (x_i - w_i)^2\right)^{-1},$

3. CORNER PEAK: $\quad f^{(3)}(\mathbf{x}) \;=\; \left(1 + \sum_{i=1}^{s} c_i x_i\right)^{-(s+1)},$

4. GAUSSIAN: $\quad f^{(4)}(\mathbf{x}) \;=\; \exp\!\left(-\sum_{i=1}^{s} c_i^2 (x_i - w_i)^2\right),$

5. CONTINUOUS: $\quad f^{(5)}(\mathbf{x}) \;=\; \exp\!\left(-\sum_{i=1}^{s} c_i |x_i - w_i|\right),$

6. DISCONTINUOUS: $\quad f^{(6)}(\mathbf{x}) \;=\; \begin{cases} 0, & \text{if } x_1 > w_1 \\ & \text{or } x_2 > w_2, \\ \exp\!\left(\sum_{i=1}^{s} c_i x_i\right), & \text{otherwise.} \end{cases}$

For convenience, all of these functions were subsequently normalized so that the true integrals over C^s equalled 1. Different test integrals were obtained by varying the parameters $\mathbf{c} = (c_1, \ldots, c_s)$ and $\mathbf{w} = (w_1, \ldots, w_s)$. Given a pre-determined sample size S, one generates S samples of a given test family in a given dimension by randomly generating S suitable pairs $\mathbf{c}$ and $\mathbf{w}$. Here the vector $\mathbf{w}$ acts as a shift parameter, so its components are chosen to be uniformly distributed in $[0, 1]$. The difficulty of the problem is controlled by the vector $\mathbf{c}$, with the difficulty of the integral usually increasing as the Euclidean norm $\|\mathbf{c}\|$ is increased. For test family j with integrand $f^{(j)}$, $1 \leqslant j \leqslant 6$, $\mathbf{c}$ is determined by using $\mathbf{c} = \gamma_j \mathbf{c}'$, where the components of $\mathbf{c}'$ are uniformly distributed in $[0, 1]$ and γ_j is such that

$$\gamma_j s^{e_j} \sum_{i=1}^{s} c_i' = h_j.$$

The numbers e_j and h_j, $1 \leqslant j \leqslant 6$, are fixed for each test family and each pre-determined level of difficulty; the precise values used in the calculations are indicated below.

For the lattice methods COPY and RANK1, we have seen that it is usual to require that the integrand be continuous and periodic. Since the integrands $f^{(j)}$, $1 \leqslant j \leqslant 4$, are smooth, it was decided to apply the periodizing transformation proposed by Sidi (1993); see Section 2.12. Thus for these integrands the transformation

$$t_i = x_i - \sin(2\pi x_i)/(2\pi), \qquad 1 \leqslant i \leqslant s,$$

was applied to each variable. Weaker transformations were used for the other two families, because those integrands are less regular. We saw in Chapter 2 that if we use the transformation

$$t = 3x^2 - 2x^3$$

on a one-dimensional function then the periodic extension of the transformed integrand usually has discontinuities in its first derivative at the integer points. Since the integrand $f^{(5)}$, though continuous, already has discontinuities in its first partial derivatives, it seemed appropriate to use this transformation on each variable for this family, on the argument that there is no point in making the regularity at the boundary better than that in the interior. On the other hand, because $f^{(6)}$ is a discontinuous function, and would still be discontinuous after applying a periodizing transformation, it seemed appropriate not to apply any transformation. This sixth test family is not really suitable for lattice rules approached in the present spirit of minimizing P_2, because the Fourier series expansion for a function in this family will not be absolutely convergent and so does not fit into the usual theoretical framework. However, we know from Chapter 8 that lattice rules can continue to have a role even for discontinuous functions. (Note, though, that for consistency we here use for $f^{(6)}$, as for every other family, the vector $\mathbf{z}$ that minimizes P_2, rather than the choice of $\mathbf{z}$ advocated in Chapter 8.) For the other two methods (that is, MCARLO and ADAPT) no transformation was applied to any of the integrands, since their theory does not require the integrands to be periodic.

In the numerical tests the sample size S was taken to be 20. Thus each run consisted of 20 test integrands from each test family. As suggested in Genz (1984), a portable random number generator (Schrage 1979) was used to obtain the vectors $\mathbf{c}'$ and $\mathbf{w}$. We restricted the dimension s to the values 5, 8, and 10. For each method tested, we used values of N approximately equal to 5×10^3, 10^4, 2×10^4, 4×10^4, 8×10^4, 1.6×10^5, 3.2×10^5, 6.4×10^5, 1.25×10^6, and 2.5×10^6, thus approximately doubling at each step. During a run, s would be fixed and results obtained for each test family using values of N as far as possible the same. (For each of COPY and RANK1, the value of N used was exactly the same for all samples in the run, but it was not possible to achieve this for MCARLO and ADAPT because of their adaptive nature. The latter two methods require a maximum permissible value of N to be specified.)

To enable us to see the differences that may occur in the four methods when the 'hardness' of the test integrands varies, runs with two differing

Table 11.1 Values of h_j for $s = 5, 8$, and 10.
$L = 1$ denotes the easier choice, $L = 2$ the harder

		$s = 5$	$s = 8$	$s = 10$
L	j	h_j	h_j	h_j
1	1	99.2	240.0	315.0
1	2	145.0	342.0	725.0
1	3	145.0	50.5	—
1	4	113.5	269.0	703.0
1	5	127.0	320.0	—
1	6	20.0	49.0	430.0
2	1	145.7	354.0	900.0
2	2	261.0	545.0	1760.0
2	3	433.0	193.0	185.0
2	4	155.0	382.0	1230.0
2	5	217.0	674.0	2040.0
2	6	90.0	240.0	1470.0

levels of difficulty were made. We shall denote the level of difficulty by L,
with $L = 1$ denoting the easier level of difficulty and $L = 2$ the harder
level. The difficulty levels were determined by making two different choices
of the parameters h_j, given in Table 11.1. All the e_j, $1 \leqslant j \leqslant 6$, were taken
to be 1.0, 1.5, and 2.0 for $s = 5$, 8, and 10 respectively.

For each test family in each run, the following five quantities were calculated. The first measures the actual accuracy of a method, the second
the apparent accuracy as judged by the error estimate. The next two assess the safety of the error estimate. Finally, to be useful an error estimate
must not only be safe, but must also give a meaningful indication of the
true error. The fifth quantity below measures the extent to which the error
estimate overestimates the true error.

(i) The median of the actual number of correct digits;

(ii) the median of the estimated number of correct digits;

(iii) the *reliability ratio*, which is the proportion of the samples in which
the actual number of correct digits is at least as great as the estimated
number of correct digits;

(iv) the *failure measure*, which is the median of the difference between the
estimated number of correct digits and the actual number of correct

digits, taken only over the samples for which the former is greater than the latter;

(v) the *error overestimation*, which is the difference between the medians of the actual and estimated number of correct digits.

The reliability ratio tells us how often the error estimate proved trustworthy. The failure measure tells us how bad the failure was on those (hopefully rare) occasions when it proved not to be trustworthy. We suggest that the reliability ratio and failure measure should both be considered when assessing the safety of an error estimate. The reliability ratio, failure measure, and error overestimation could not be calculated when testing the rank-1 rules (RANK1), since no error estimate was available.

11.3   Numerical results

For reasons of space, we cannot give the full details of the numerical results. However, for each family, we do give graphs comparing the accuracy of the methods for each dimension and level of difficulty tested. These are given by Figures 11.1–11.34, which are plots of the median number of correct digits for each run against $\log_{10}(N)$.

For the various quality measures of the error estimate, there is space only for short summaries. To explain how these are constructed, we give for test family 1, $s = 5$, and $L = 1$ a full table of the reliability ratios for each run (in Table 11.2), a full table of the corresponding failure measures (in Table 11.3), and a full table of the error overestimation, the differences between the medians of the actual and estimated number of correct digits (in Table 11.4). The table of reliability ratios is then summarized by taking the average of the reliability ratios over the ten runs for the ten values of N. Thus we end up with the values given in the first row of Table 11.5.

To summarize the failure measures as given by a table such as Table 11.3, an average of the failure measures is also calculated, but only over the failure measures which are nonzero. If the reliability ratios for all ten runs are 1, then all the failure measures are 0. This situation is reflected in the summary table by the entry —. The first row of Table 11.6 contains the summary of Table 11.3. In a similar way, to summarize the error overestimation the average is calculated over the separate error overestimations.

Table 11.2 Reliability results for $s = 5$, $L = 1$,
and family 1

N (approx.)	COPY	MCARLO	ADAPT
5000	1.00	0.95	1.00
10000	1.00	1.00	1.00
20000	1.00	0.90	1.00
40000	1.00	1.00	1.00
80000	1.00	1.00	1.00
160000	1.00	1.00	1.00
320000	1.00	0.95	1.00
640000	1.00	0.95	1.00
1250000	1.00	0.90	1.00
2500000	1.00	0.90	1.00

Table 11.3 Failure measure results for $s = 5$,
$L = 1$, and family 1

N (approx.)	COPY	MCARLO	ADAPT
5000	—	0.03	—
10000	—	—	—
20000	—	0.08	—
40000	—	—	—
80000	—	—	—
160000	—	—	—
320000	—	0.08	—
640000	—	0.04	—
1250000	—	0.02	—
2500000	—	0.04	—

Table 11.4 Error overestimation for $s = 5$, $L = 1$, and family 1

N (approx.)	COPY	MCARLO	ADAPT
5000	1.87	0.04	2.80
10000	2.30	0.59	3.69
20000	3.10	0.45	3.70
40000	2.23	0.49	4.06
80000	2.25	0.49	4.62
160000	2.06	0.31	4.91
320000	1.47	0.83	4.91
640000	2.35	0.46	5.44
1250000	1.16	0.30	5.62
2500000	2.57	0.36	5.52

Family 1 (OSCILLATORY): The results for test family 1, the easiest of the six families, are given in Figures 11.1–11.6 and Tables 11.5–11.7. In terms of accuracy and reliability of the error estimate, it is clear that for this family MCARLO is not competitive. For accuracy, ADAPT seems to have the edge over the other three methods for family 1, with the advantage increasing as s increases. However, the slopes of the lines for COPY indicate that for sufficiently large N, COPY could actually be more accurate than ADAPT. This suggests that the rate of convergence of COPY is higher than that of ADAPT. At the highest values of s and N one can see that COPY is also more accurate than RANK1. For reliability, ADAPT is marginally better than COPY. The failure measure results for both COPY and MCARLO are reasonable. The error overestimation for MCARLO is significantly less than for COPY, which in turn is significantly less than for ADAPT.

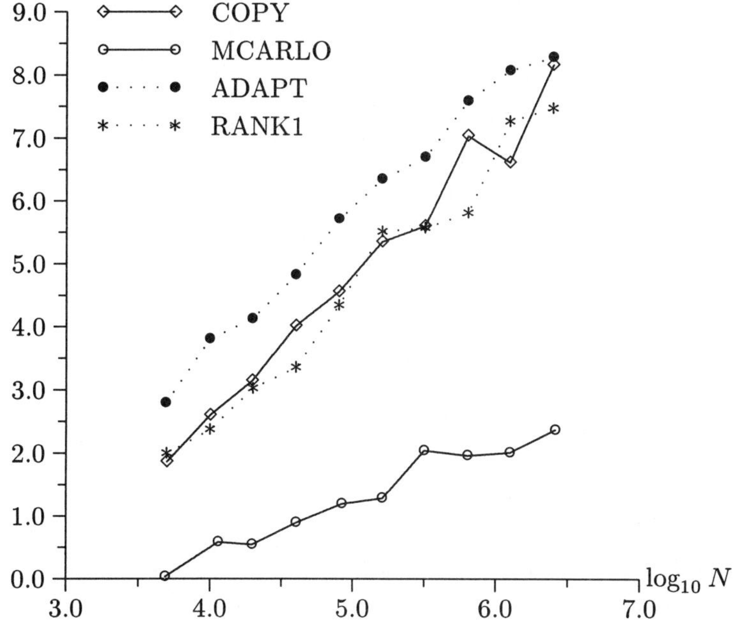

Fig. 11.1 Median number of correct digits for $s = 5$, $L = 1$, and family 1.

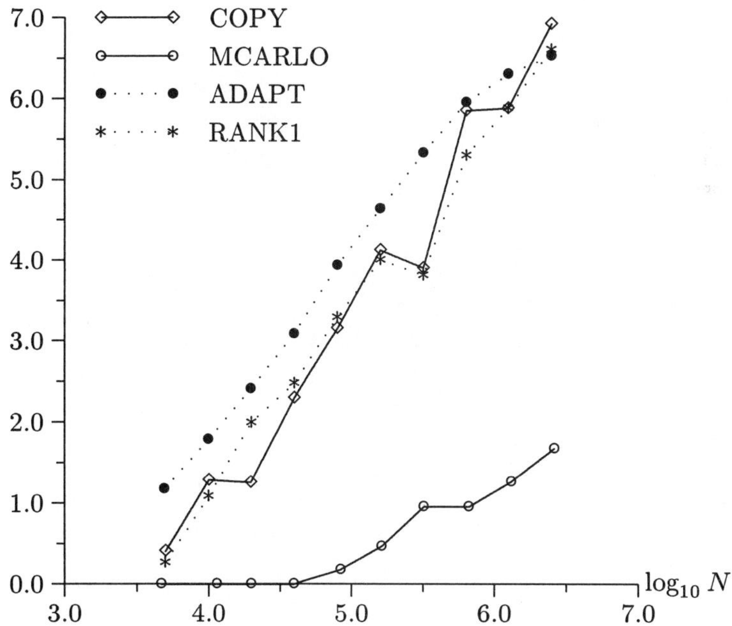

Fig. 11.2 Median number of correct digits for $s = 5$, $L = 2$, and family 1.

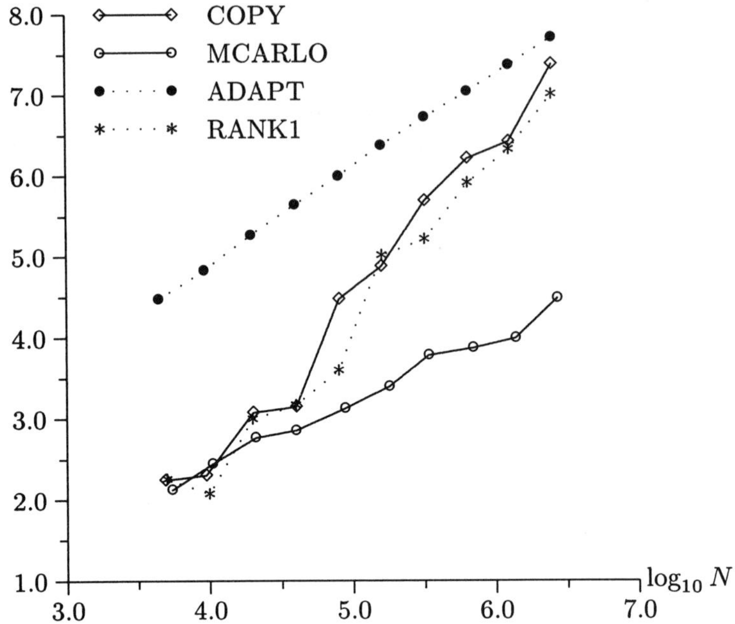

Fig. 11.3 Median number of correct digits for $s = 8$, $L = 1$, and family 1.

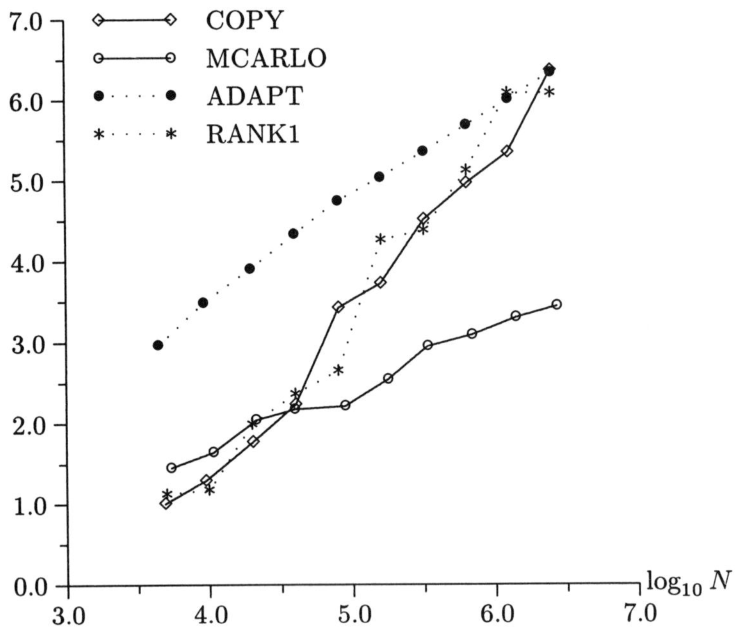

Fig. 11.4 Median number of correct digits for $s = 8$, $L = 2$, and family 1.

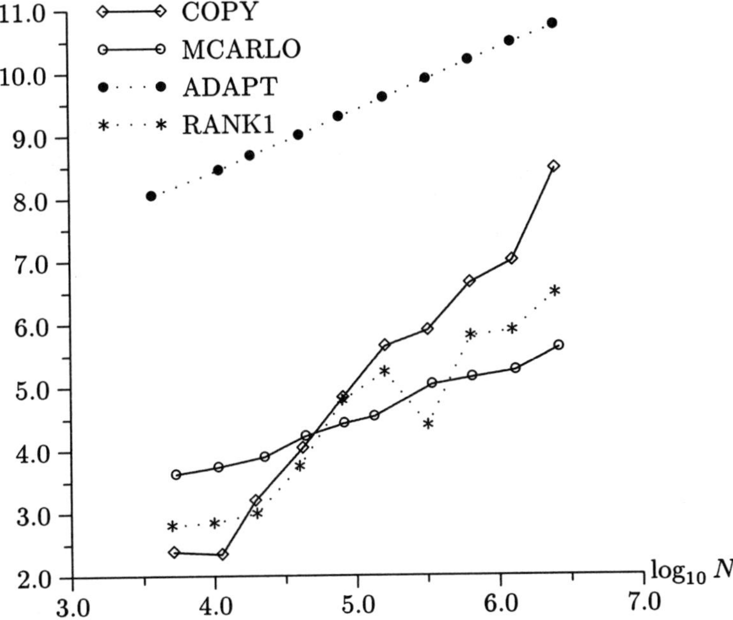

Fig. 11.5 Median number of correct digits for $s = 10$, $L = 1$, and family 1.

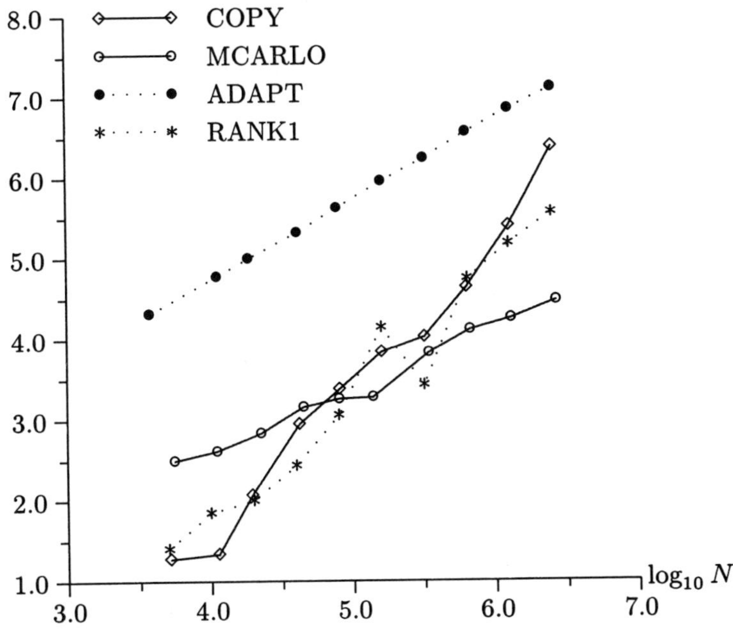

Fig. 11.6 Median number of correct digits for $s = 10$, $L = 2$, and family 1.

Table 11.5 Summary of reliability for family 1

s	L	COPY	MCARLO	ADAPT
5	1	1.000	0.955	1.000
5	2	0.980	0.925	1.000
8	1	1.000	0.960	1.000
8	2	0.985	0.965	1.000
10	1	1.000	0.965	1.000
10	2	1.000	0.965	1.000

Table 11.6 Summary of failure measure for family 1

s	L	COPY	MCARLO	ADAPT
5	1	—	0.048	—
5	2	0.550	0.051	—
8	1	—	0.058	—
8	2	0.060	0.073	—
10	1	—	0.063	—
10	2	—	0.087	—

Table 11.7 Summary of error overestimation for family 1

s	L	COPY	MCARLO	ADAPT
5	1	2.136	0.432	4.527
5	2	1.646	0.319	3.671
8	1	1.795	0.368	2.714
8	2	1.480	0.448	2.774
10	1	1.671	0.442	3.163
10	2	1.230	0.380	2.354

Family 2 (PRODUCT PEAK): The results for test family 2 are given in Figures 11.7–11.12 and Tables 11.8–11.10. In terms of accuracy and reliability of the error estimate, it is clear that MCARLO is not competitive with the other three methods (except against RANK1 for $s = 10$ and $L = 2$). For accuracy, COPY perhaps has the edge over the other three methods (for $s = 10$ and $L = 2$ the slope of the line for COPY in Figure 11.12 indicates that COPY would be more accurate than ADAPT if a value of N somewhat larger than the maximum of about 2.5×10^6 was used). ADAPT is doing relatively better at the higher level of difficulty and higher values of s. For reliability, ADAPT and COPY are about the same. The failure measure results for MCARLO are better than for ADAPT, which in turn are slightly better than COPY. The error overestimation for MCARLO is significantly less than for the other two methods.

 Comparisons with other methods

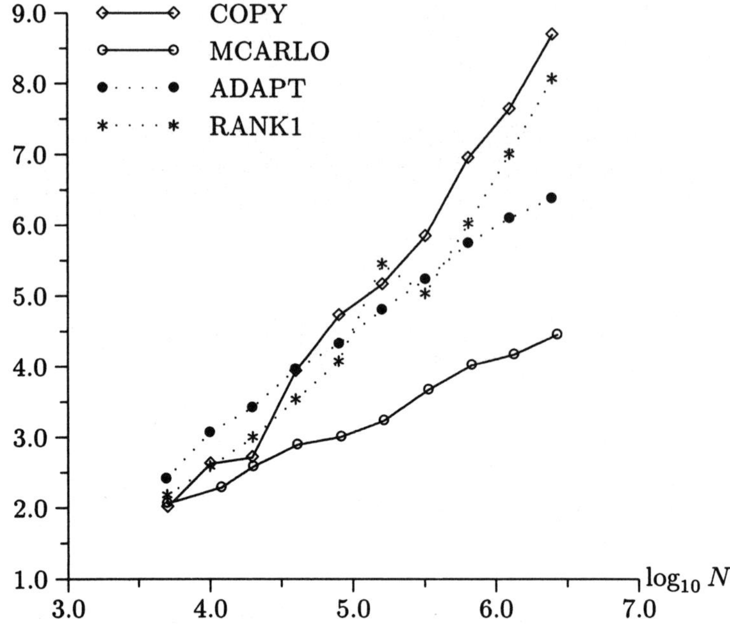

Fig. 11.7 Median number of correct digits for $s = 5$, $L = 1$, and family 2.

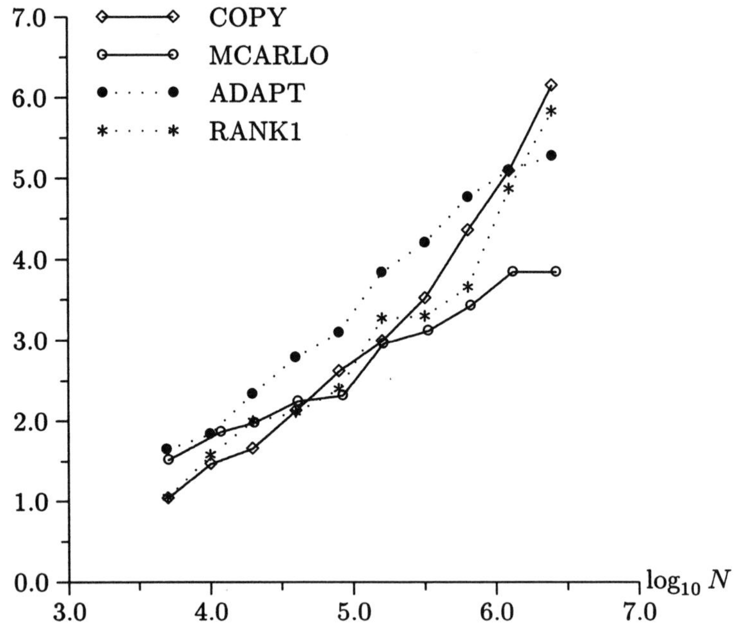

Fig. 11.8 Median number of correct digits for $s = 5$, $L = 2$, and family 2.

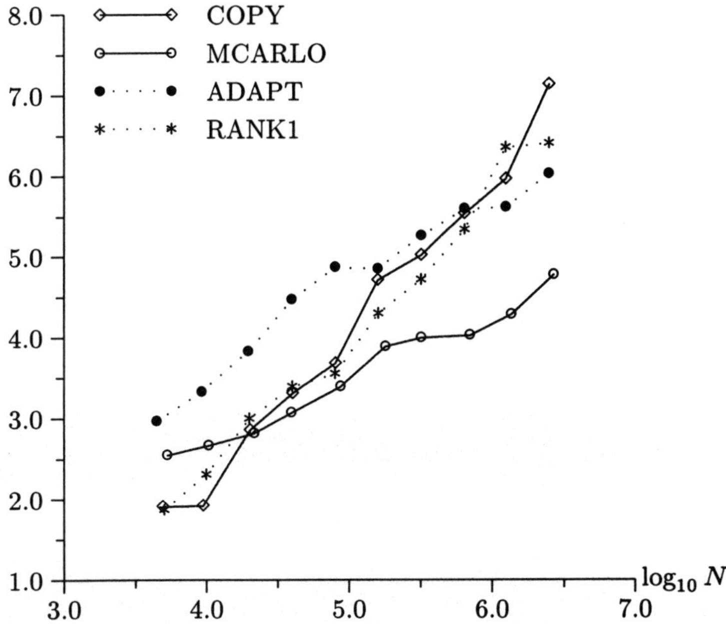

Fig. 11.9 Median number of correct digits for $s = 8$, $L = 1$, and family 2.

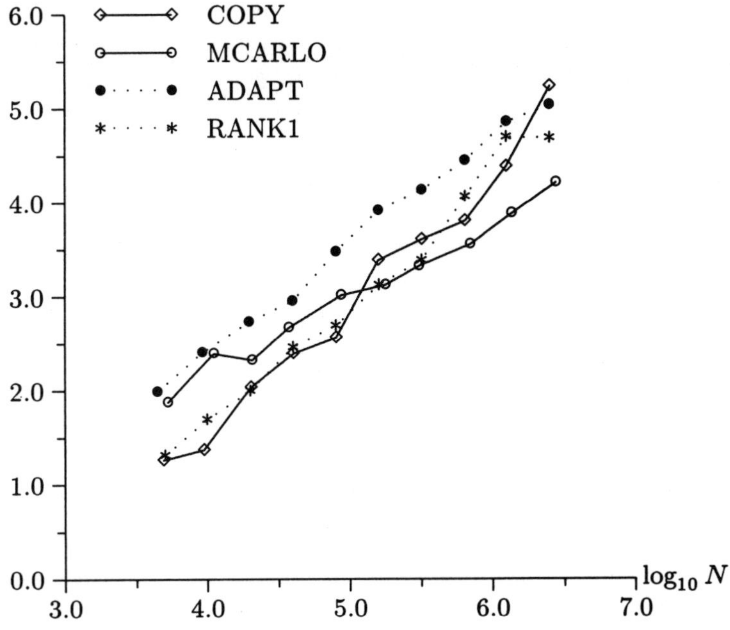

Fig. 11.10 Median number of correct digits for $s = 8$, $L = 2$, and family 2.

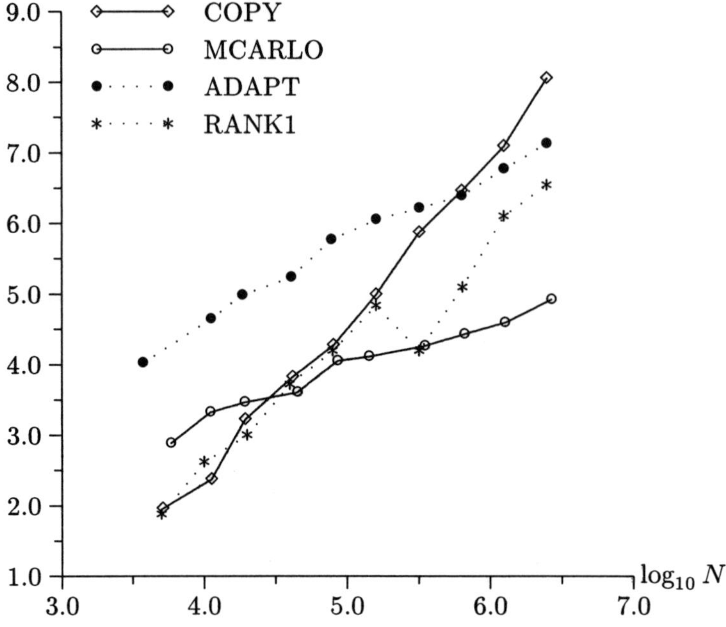

Fig. 11.11 Median number of correct digits for $s = 10$, $L = 1$, and family 2.

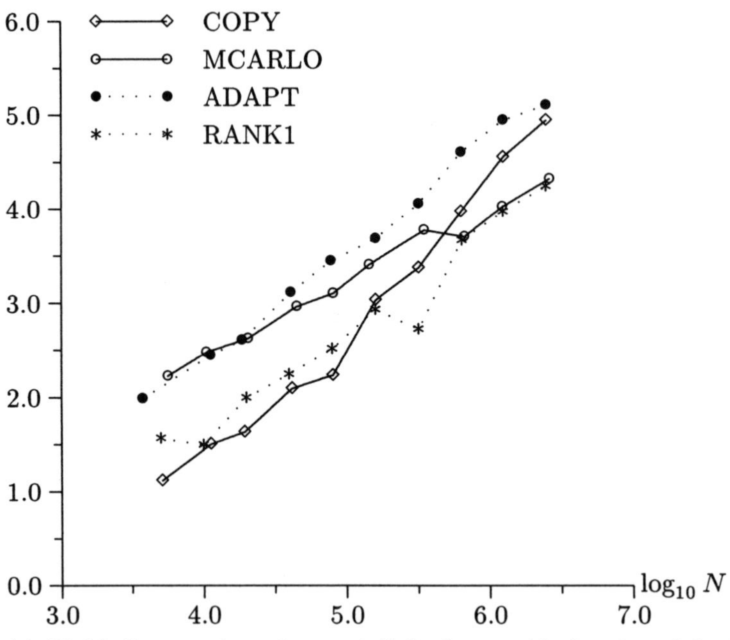

Fig. 11.12 Median number of correct digits for $s = 10$, $L = 2$, and family 2.

Table 11.8 Summary of reliability for family 2

s	L	COPY	MCARLO	ADAPT
5	1	0.995	0.960	0.995
5	2	0.935	0.945	0.945
8	1	1.000	0.930	0.995
8	2	0.980	0.960	1.000
10	1	1.000	0.935	0.990
10	2	0.985	0.960	1.000

Table 11.9 Summary of failure measure for family 2

s	L	COPY	MCARLO	ADAPT
5	1	0.320	0.077	0.170
5	2	0.340	0.094	0.298
8	1	—	0.085	0.050
8	2	0.075	0.077	—
10	1	—	0.087	0.120
10	2	0.170	0.098	—

Table 11.10 Summary of error overestimation for family 2

s	L	COPY	MCARLO	ADAPT
5	1	1.765	0.449	1.374
5	2	1.079	0.524	1.022
8	1	1.606	0.481	1.716
8	2	1.100	0.469	1.277
10	1	1.999	0.435	1.820
10	2	1.081	0.476	1.405

Family 3 (CORNER PEAK): The results for test family 3 are given in Figures 11.13–11.17 and Tables 11.11–11.13. In terms of accuracy and reliability of the error estimate, it is clear that MCARLO is not generally competitive with the other three methods (except for some of the smallest values of N). For accuracy, ADAPT looks best at $s = 5$, but at the two other (higher) values of s COPY takes over at the largest values of N. COPY also tends to be better than RANK1 at the highest values of N. For reliability, ADAPT has the edge over COPY. The failure measure results for MCARLO are slightly better than for COPY, but the latter are still reasonable. Error overestimation for MCARLO is significantly less than for COPY, which in turn is less than for ADAPT.

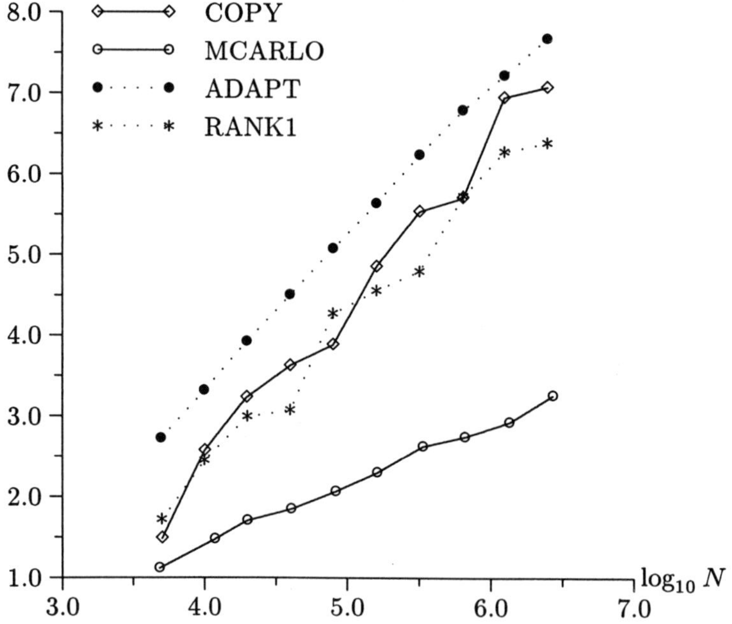

Fig. 11.13 Median number of correct digits for $s = 5$, $L = 1$, and family 3.

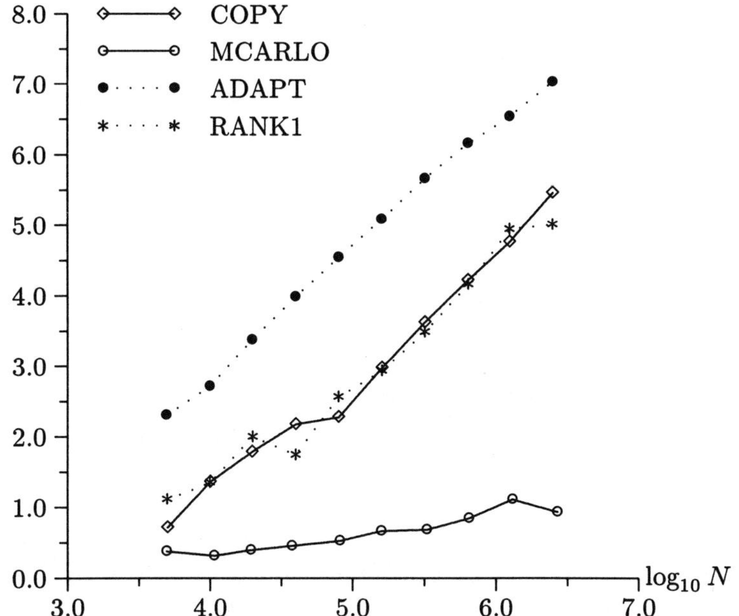

Fig. 11.14 Median number of correct digits for $s = 5$, $L = 2$, and family 3.

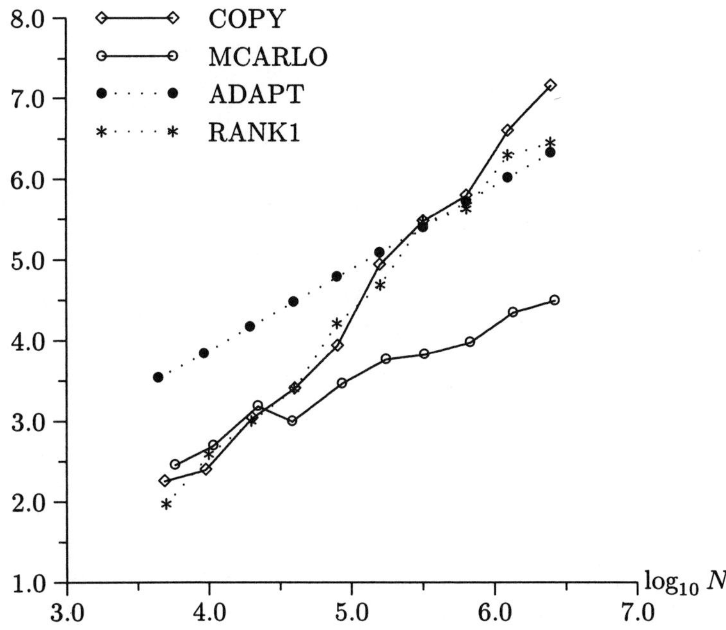

Fig. 11.15 Median number of correct digits for $s = 8$, $L = 1$, and family 3.

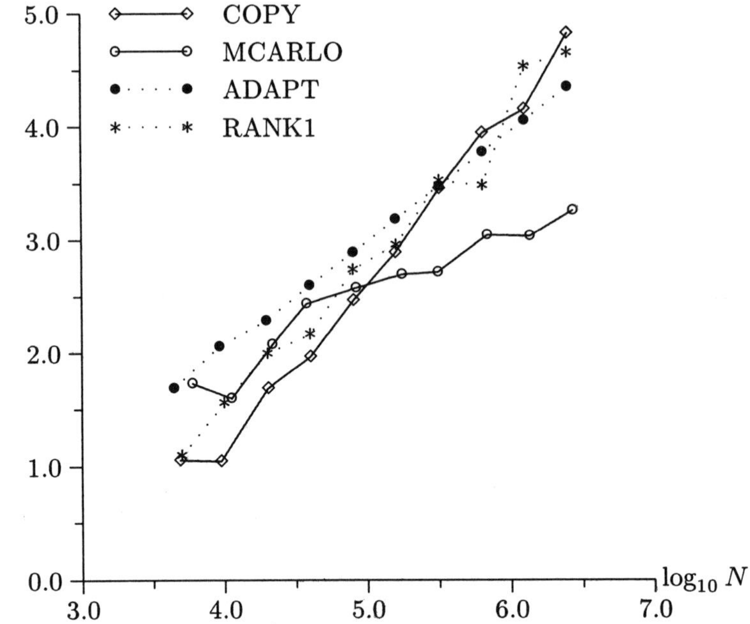

Fig. 11.16 Median number of correct digits for $s = 8$, $L = 2$, and family 3.

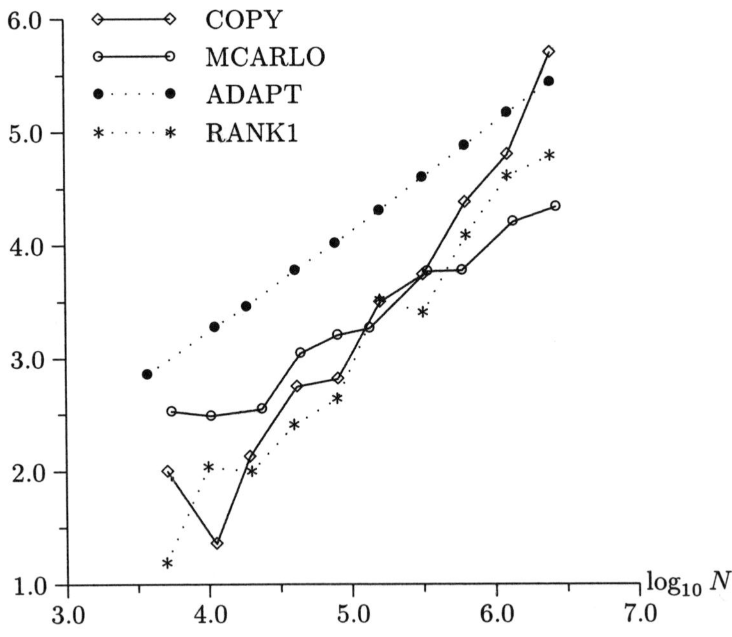

Fig. 11.17 Median number of correct digits for $s = 10$, $L = 2$, and family 3.

Table 11.11 Summary of reliability for family 3

s	L	COPY	MCARLO	ADAPT
5	1	0.985	0.815	1.000
5	2	0.910	0.580	1.000
8	1	1.000	0.945	1.000
8	2	0.880	0.925	1.000
10	2	0.980	0.935	1.000

Table 11.12 Summary of failure measure for family 3

s	L	COPY	MCARLO	ADAPT
5	1	0.280	0.128	—
5	2	0.125	0.261	—
8	1	—	0.083	—
8	2	0.120	0.086	—
10	2	0.055	0.058	—

Table 11.13 Summary of error overestimation for family 3

s	L	COPY	MCARLO	ADAPT
5	1	1.592	0.331	2.122
5	2	1.049	0.087	1.985
8	1	1.503	0.473	1.806
8	2	0.868	0.444	1.288
10	2	1.130	0.467	1.599

Family 4 (GAUSSIAN): The results for test family 4 are given in Figures 11.18–11.23 and Tables 11.14–11.16. Again, as for the other three test families already considered, MCARLO cannot really be recommended. For accuracy, ADAPT performs well. At the highest values of N, COPY looks best for $L = 1$ while ADAPT looks best for $L = 2$, and the slopes of the lines for COPY when $L = 2$ indicate that for N sufficiently large COPY would be more accurate than ADAPT. For reliability, ADAPT has a slight edge over COPY. The failure measure results for MCARLO are better than for COPY though these latter ones are still reasonable. The error overestimation for MCARLO is less than for the other two methods. For the two higher values of s, error overestimation for ADAPT is worse than for COPY.

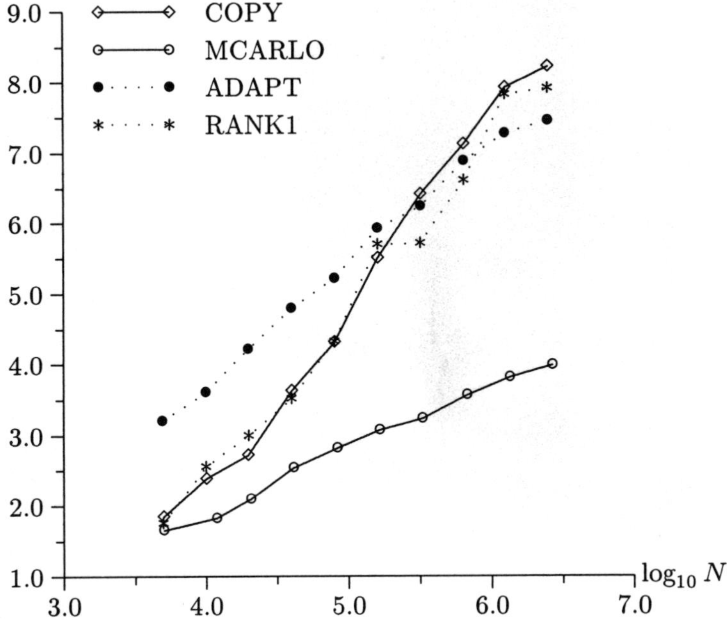

Fig. 11.18 Median number of correct digits for $s = 5$, $L = 1$, and family 4.

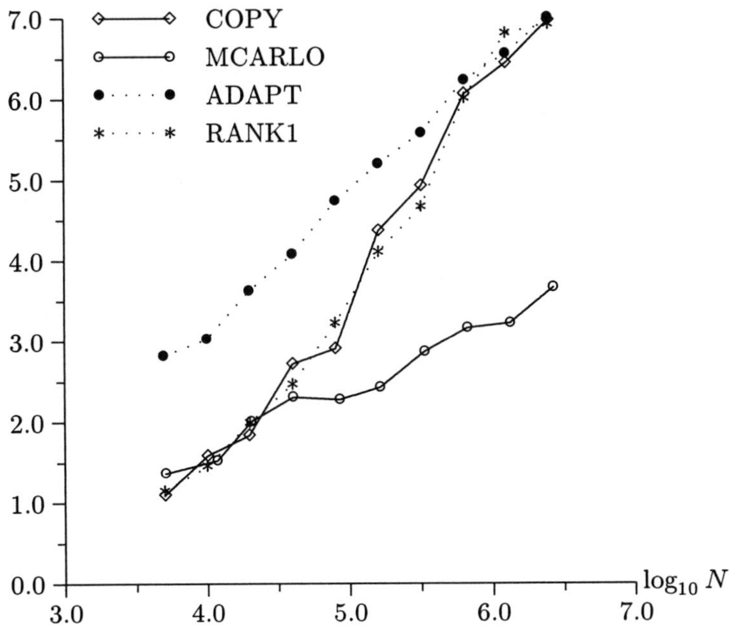

Fig. 11.19 Median number of correct digits for $s = 5$, $L = 2$, and family 4.

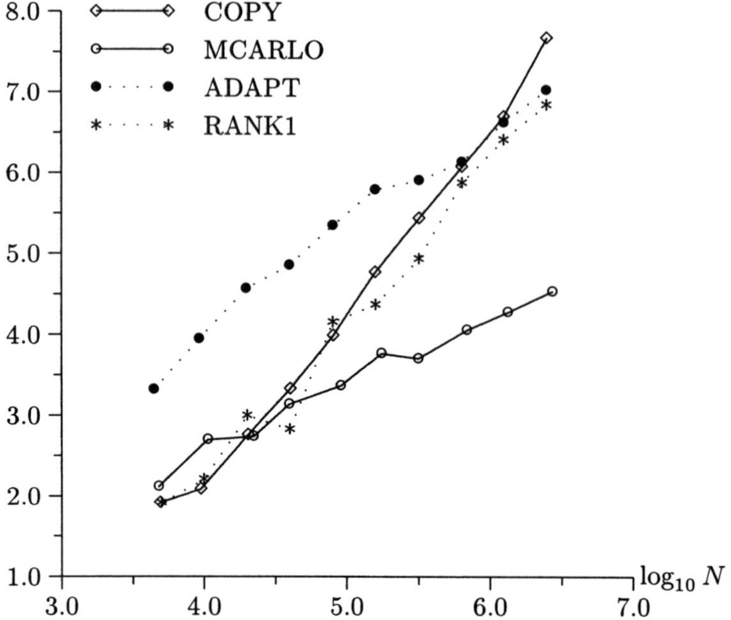

Fig. 11.20 Median number of correct digits for $s = 8$, $L = 1$, and family 4.

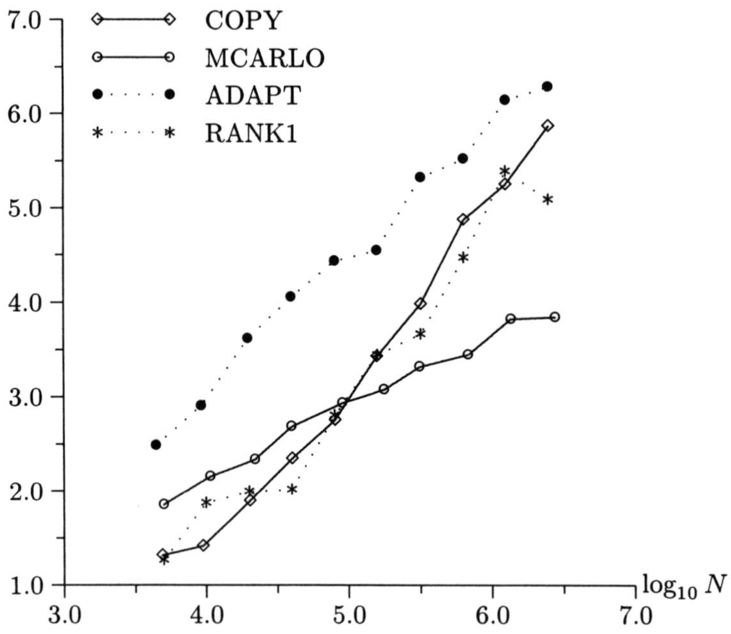

Fig. 11.21 Median number of correct digits for $s = 8$, $L = 2$, and family 4.

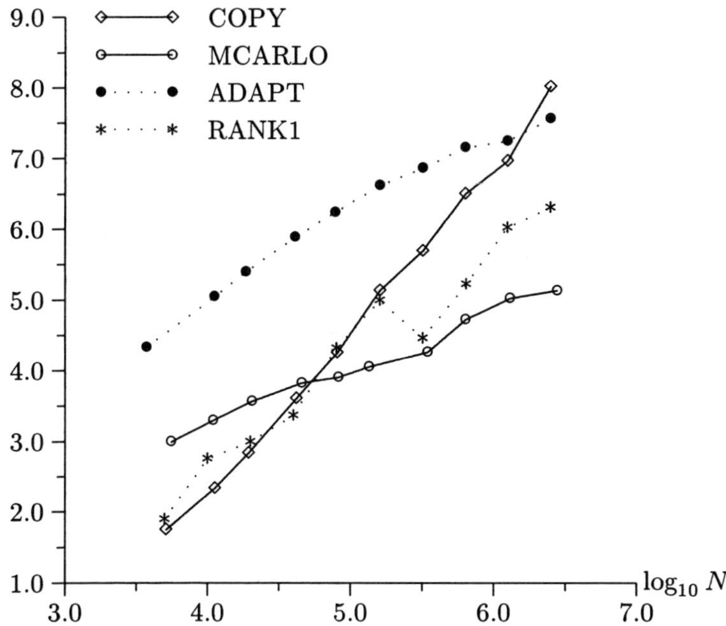

Fig. 11.22 Median number of correct digits for $s = 10$, $L = 1$, and family 4.

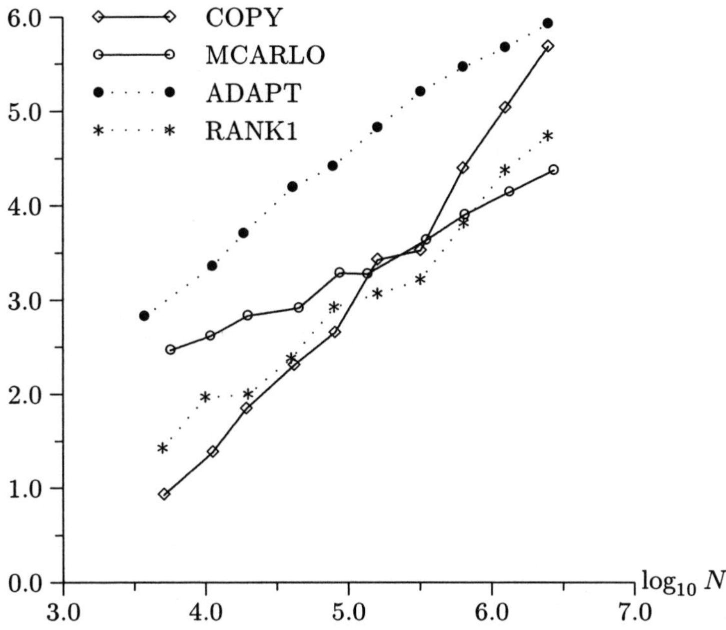

Fig. 11.23 Median number of correct digits for $s = 10$, $L = 2$, and family 4.

Table 11.14 Summary of reliability for family 4

s	L	COPY	MCARLO	ADAPT
5	1	0.990	0.940	1.000
5	2	0.975	0.920	1.000
8	1	1.000	0.945	1.000
8	2	0.995	0.940	1.000
10	1	1.000	0.970	1.000
10	2	0.995	0.955	1.000

Table 11.15 Summary of failure measure for family 4

s	L	COPY	MCARLO	ADAPT
5	1	0.250	0.086	—
5	2	0.200	0.090	—
8	1	—	0.071	—
8	2	0.380	0.065	—
10	1	—	0.094	—
10	2	0.070	0.038	—

Table 11.16 Summary of error overestimation for family 4

s	L	COPY	MCARLO	ADAPT
5	1	1.908	0.475	1.734
5	2	1.500	0.501	1.469
8	1	1.773	0.435	1.957
8	2	1.340	0.463	1.818
10	1	1.937	0.475	2.448
10	2	1.246	0.480	1.960

Family 5 (CONTINUOUS): The results for test family 5 are given in Figures 11.24–11.28 and Tables 11.17–11.19. For this test family, ADAPT is not doing well in terms of accuracy against the other three methods. This is perhaps a surprise, in view of the fact that the first derivative discontinuities in this family all occur on planes perpendicular to the coordinate axes, a situation that would seem relatively favourable to ADAPT. The best method from the point of accuracy looks like COPY, though it is only marginally better than RANK1. For reliability, COPY has a slight edge over MCARLO, while ADAPT is significantly worse than either of these two methods (especially at the highest values of s). The failure measure results for MCARLO are better than those for COPY or ADAPT, though the latter are still reasonable. Overall, error overestimation for MCARLO is somewhat less than for ADAPT, which in turn is less than for COPY, but even COPY overestimates the error by less than an order of magnitude.

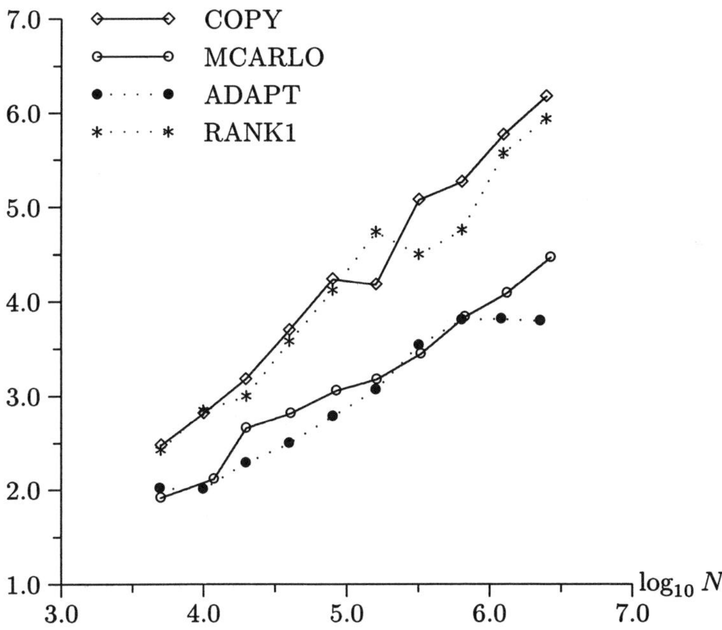

Fig. 11.24 Median number of correct digits for $s = 5$, $L = 1$, and family 5.

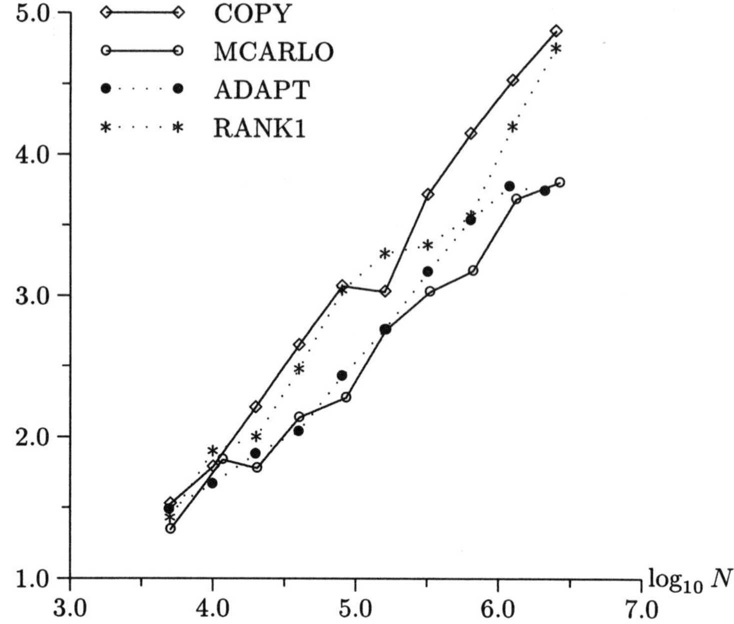

Fig. 11.25 Median number of correct digits for $s = 5$, $L = 2$, and family 5.

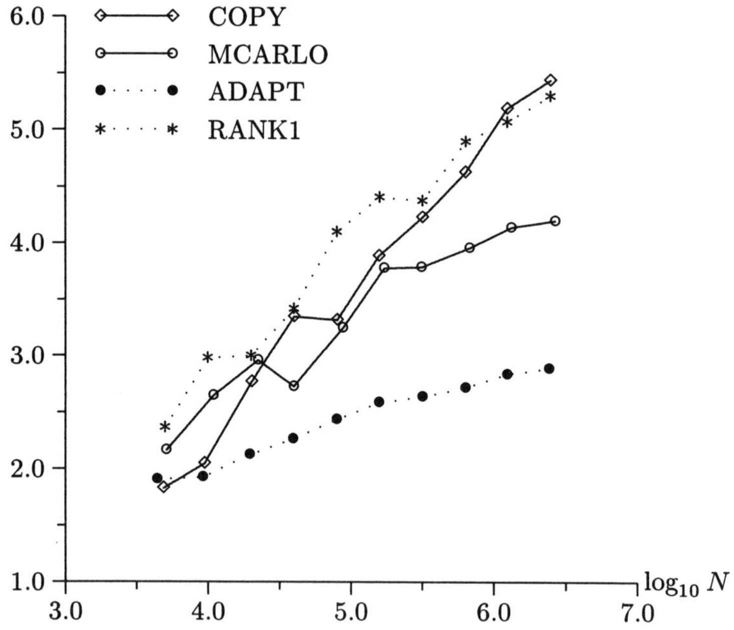

Fig. 11.26 Median number of correct digits for $s = 8$, $L = 1$, and family 5.

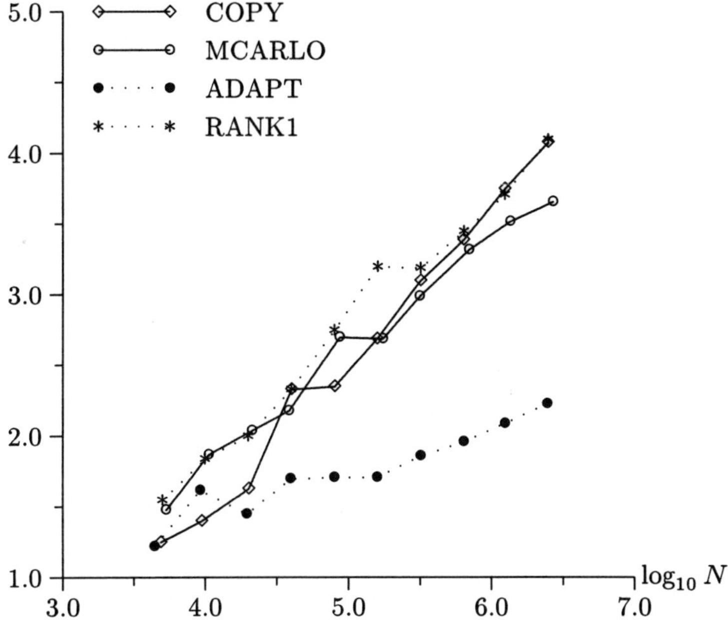

Fig. 11.27 Median number of correct digits for $s = 8$, $L = 2$, and family 5.

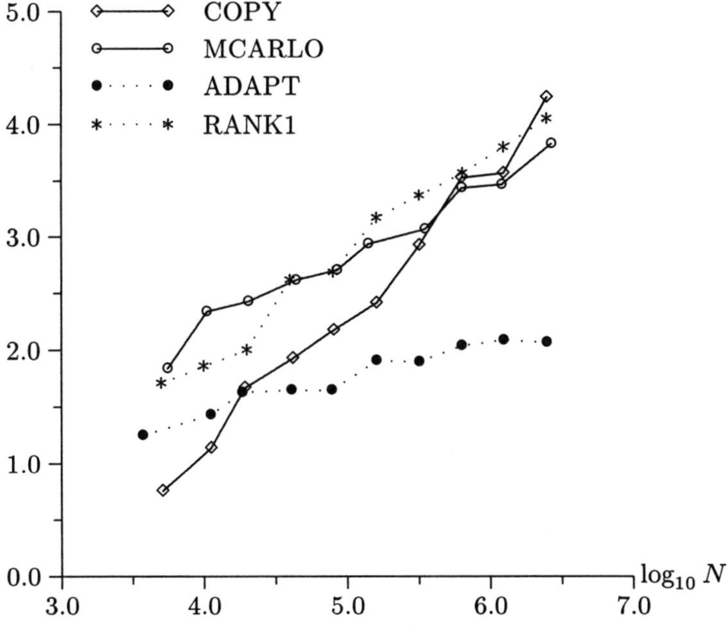

Fig. 11.28 Median number of correct digits for $s = 10$, $L = 2$, and family 5.

Table 11.17 Summary of reliability for family 5

s	L	COPY	MCARLO	ADAPT
5	1	1.000	0.950	0.935
5	2	0.985	0.940	0.960
8	1	0.990	0.965	0.815
8	2	0.975	0.955	0.845
10	2	0.970	0.960	0.815

Table 11.18 Summary of failure measure for family 5

s	L	COPY	MCARLO	ADAPT
5	1	—	0.077	0.381
5	2	0.200	0.096	0.177
8	1	0.060	0.058	0.218
8	2	0.237	0.099	0.148
10	2	0.070	0.095	0.267

Table 11.19 Summary of error overestimation for family 5

s	L	COPY	MCARLO	ADAPT
5	1	1.175	0.473	0.628
5	2	0.967	0.476	0.780
8	1	0.986	0.494	0.510
8	2	0.731	0.501	0.444
10	2	0.686	0.465	0.428

Family 6 (DISCONTINUOUS): The results for test family 6 are given in Figures 11.29–11.34 and Tables 11.20–11.22. As has already been pointed out, this test family is not really suitable for COPY or RANK1. All four methods are having difficulty with this test family. RANK1 appears to be the best of the four methods at $L = 1$, while ADAPT appears the best for $L = 2$. However, the slope of the lines for ADAPT does not inspire confidence about the results obtained if one were to use larger values of N. In terms of reliability, MCARLO is far superior to COPY and ADAPT. The reliability results for ADAPT appear to be better than those for COPY at the large values of s. However, the failure measure results for ADAPT look rather bad compared to those for COPY and MCARLO, with the ADAPT algorithm often underestimating the true error by many orders of magnitude. On the other hand, the failure measures for MCARLO are somewhat better than those for COPY. The error overestimation for COPY is somewhat better than for the other two methods, but in all cases the error overestimation is reasonable. Overall, for this test family, MCARLO seems to be the most appropriate method to use of all the methods considered here.

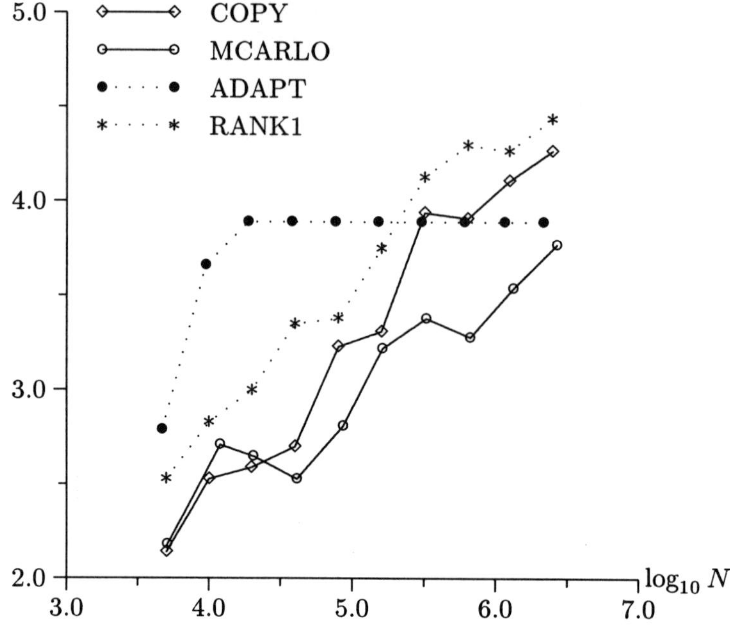

Fig. 11.29 Median number of correct digits for $s = 5$, $L = 1$, and family 6.

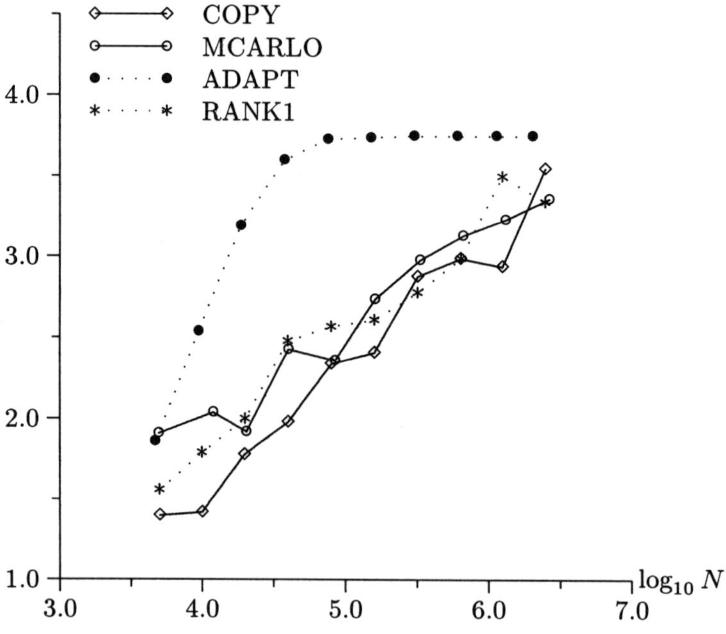

Fig. 11.30 Median number of correct digits for $s = 5$, $L = 2$, and family 6.

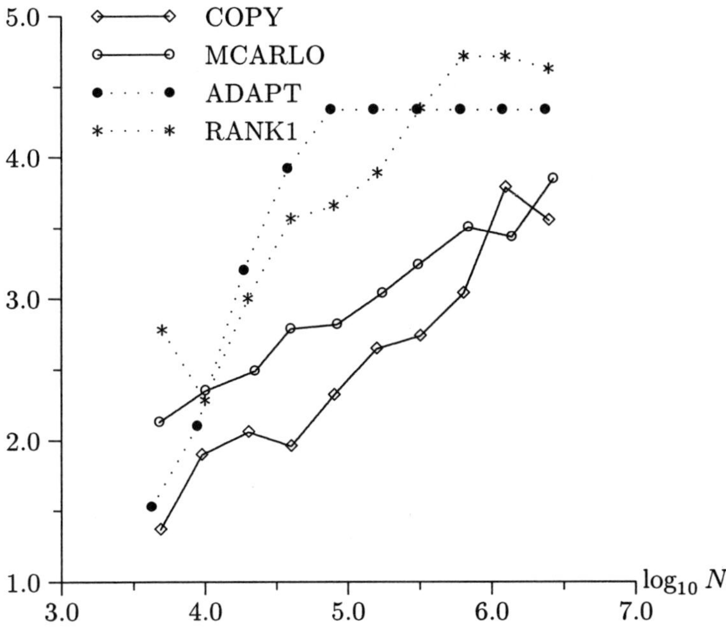

Fig. 11.31 Median number of correct digits for $s = 8$, $L = 1$, and family 6.

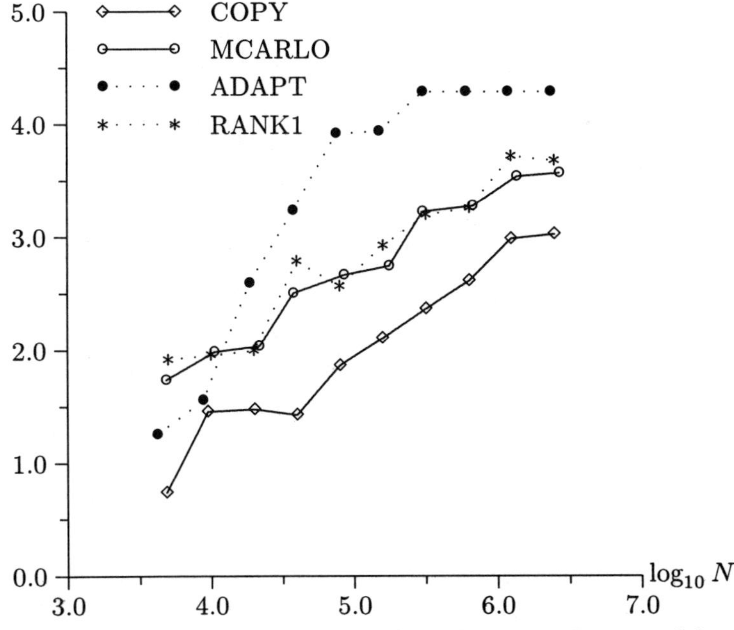

Fig. 11.32 Median number of correct digits for $s = 8$, $L = 2$, and family 6.

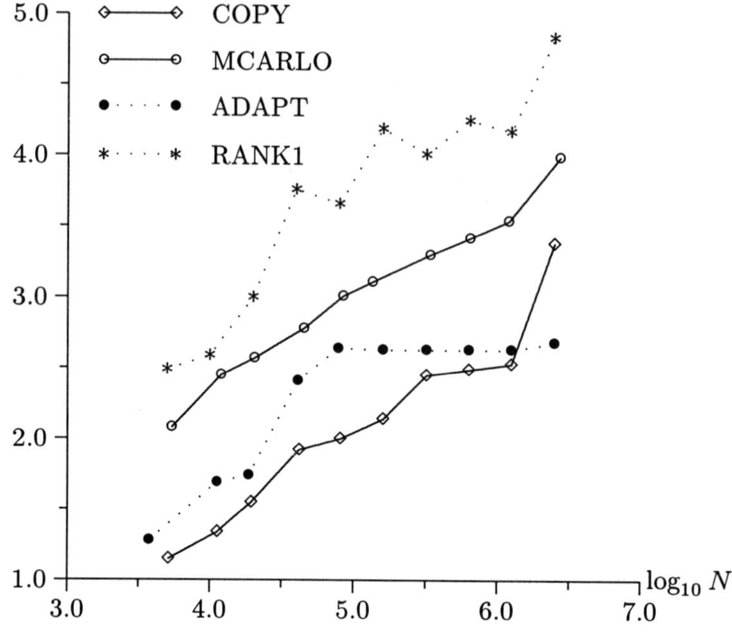

Fig. 11.33 Median number of correct digits for $s = 10$, $L = 1$, and family 6.

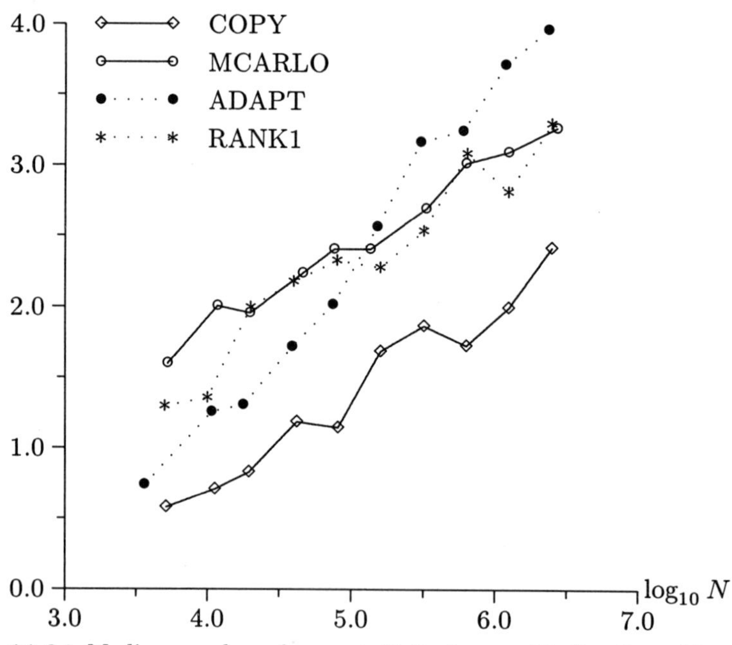

Fig. 11.34 Median number of correct digits for $s = 10$, $L = 2$, and family 6.

Table 11.20 Summary of reliability for family 6

s	L	COPY	MCARLO	ADAPT
5	1	0.695	0.950	0.365
5	2	0.830	0.960	0.425
8	1	0.560	0.945	0.545
8	2	0.675	0.945	0.700
10	1	0.450	0.975	0.600
10	2	0.560	0.910	0.870

Table 11.21 Summary of failure measure for family 6

s	L	COPY	MCARLO	ADAPT
5	1	0.205	0.064	4.231
5	2	0.146	0.049	1.738
8	1	0.282	0.092	3.377
8	2	0.196	0.127	0.441
10	1	0.234	0.102	2.205
10	2	0.235	0.076	5.810

Table 11.22 Summary of error overestimation for family 6

s	L	COPY	MCARLO	ADAPT
5	1	0.320	0.574	0.380
5	2	0.358	0.582	0.057
8	1	0.213	0.483	0.480
8	2	0.246	0.516	0.361
10	1	0.163	0.494	0.540
10	2	0.242	0.443	0.544

11.4 Conclusions

The numerical results given in the previous section indicate that for all the test families except test family 6 (in which the integrands are discontinuous), the two most successful methods are generally COPY (the lattice rule algorithm of Section 10.4) and ADAPT (the adaptive method of Van Dooren and De Ridder 1976). In terms of accuracy, ADAPT was better in some test families and COPY better in others. However, what is striking is the slope of the lines for COPY. For the figures in which ADAPT appeared to be more accurate than COPY, the indications are that for large enough N (larger than the maximum of 2.5×10^6 used), COPY would actually be more accurate than ADAPT. This suggests that the rate of convergence for lattice rules which are 2^s copies of rank-1 rules is greater than that of ADAPT, for N large enough. This conclusion is supported by the numerical results of Beckers and Haegemans (1992*b*), in which they used COPY and ADAPT to compute multivariate normal integrals (similar to the Gaussian integrals in test family 4).

Overall, the reliability results indicate that the error estimate of ADAPT is slightly more reliable than COPY, except for test family 5. In contrast, the results of Beckers and Haegemans (1992*b*) indicated that the error estimate from ADAPT was not as reliable as the one from COPY. The failure measures for both COPY and ADAPT were quite reasonable, except for test family 6 in which ADAPT very seriously underestimated the error.

The method RANK1 is generally of comparable accuracy to COPY, but has the serious drawback that in this version no error estimate is available. The procedure of error estimation by randomization, not tested here, is generally reliable, but has the overwhelming defect of increasing the number of function evaluations by (typically) a factor of four—see Section 4.6.

Finally, a word of warning to the reader intending to make serious use of high-dimensional quadrature rules: no set of numerical experiments can hope to reflect fairly the conditions that one is likely to encounter in one's own problems. The present tests are limited to just six families, as described in Section 11.2 and even these are open to criticism. For example, in the DISCONTINUOUS family all the functions have discontinuities only on planes, and not just any planes, but in fact only planes that are perpendicular to one of the coordinate axes. It may be that this location of the planes is advantageous to the adaptive method, in that the difficult region can be covered easily by rectangular solids, whereas a sloping or curved surface of discontinuity cannot. A similar comment applies to the CONTINUOUS family, which has discontinuities of first derivatives on planes parallel to the axes. On the other hand, in the present experiments we have introduced a bias of a different kind, by restricting the difficulty of the problems so that at least some of the methods can achieve reasonable accuracy. It is a regrettable truth that real-world problems are not so obliging, and indeed can easily be intractable to all known methods! The moral is to treat all numerical tests with a degree of caution, to use more than one method if possible, and to treat all numerical results with a degree of scepticism, unless they have guaranteed error bounds (and in the quadrature world such error bounds are rare indeed); and, finally, to press workers in the quadrature industry to produce both improved methods and improved test families.

With this caveat, the numerical results presented here suggest that 2^s copy lattice rules as implemented in COPY make a useful addition to the armoury of methods available for multiple integration. In the future we hope that these and other lattice methods will have a useful role to play in the practical evaluation of multidimensional integrands.

Appendix A

Tables of z for 2^s copy rules

Here we give tables of $\mathbf{z}$ vectors suitable for use in the algorithm of Section 10.4 (with $n = 2$). These are obtained by finding the value of ℓ in

$$\mathbf{z}(\ell) = (1, \ell, \ell^2 \bmod m, \ldots, \ell^{s-1} \bmod m), \qquad 1 \leqslant \ell \leqslant \lfloor m/2 \rfloor,$$

that minimizes $P_2(Q_s)$ with $n = 2$, and then carrying out a permutation procedure. Further details are given in Section 10.5.

Table A.1 Recommended choices of $\mathbf{z}$ for $s = 2$

m	$2^s m$	ℓ	$\mathbf{z}$	$P_2(Q_s)$
1249	4996	512	$(512, 1)$	$0.939(-5)$
2503	10012	672	$(672, 1)$	$0.247(-5)$
5003	20012	1850	$(1, 1850)$	$0.632(-6)$
10007	40028	3822	$(1, 3822)$	$0.175(-6)$
20011	80044	6103	$(6103, 1)$	$0.470(-7)$
40009	160036	15152	$(1, 15152)$	$0.124(-7)$
80021	320084	30954	$(30954, 1)$	$0.328(-8)$

Table A.2 Recommended choices of $\mathbf{z}$ for $s = 3$

m	$2^s m$	ℓ	$\mathbf{z}$	$P_2(Q_s)$
619	4952	178	$(178, 1, 115)$	$0.347(-3)$
1249	9992	136	$(1010, 136, 1)$	$0.107(-3)$
2503	20024	652	$(2097, 1, 652)$	$0.311(-4)$
5003	40024	1476	$(1, 2271, 1476)$	$0.908(-5)$
10007	80056	2325	$(2325, 1, 1845)$	$0.286(-5)$
20011	160088	2894	$(2894, 10638, 1)$	$0.832(-6)$
40009	320072	8789	$(29151, 8789, 1)$	$0.240(-6)$

Table A.3 Recommended choices of **z** for $s = 4$

m	$2^s m$	ℓ	**z**	$P_2(Q_s)$
313	5008	51	$(51, 1, 97, 252)$	$0.545(-2)$
619	9904	73	$(1, 285, 377, 73)$	$0.198(-2)$
1249	19984	197	$(90, 1, 244, 197)$	$0.647(-3)$
2503	40048	792	$(151, 1514, 792, 1)$	$0.207(-3)$
5003	80048	792	$(1889, 1, 792, 191)$	$0.690(-4)$
10007	160112	1206	$(1, 1206, 3421, 2842)$	$0.231(-4)$
20011	320176	8455	$(7733, 8455, 1, 6578)$	$0.778(-5)$

Table A.4 Recommended choices of **z** for $s = 5$

m	$2^s m$	ℓ	**z**	$P_2(Q_s)$
157	5024	18	$(100, 18, 23, 10, 1)$	$0.504(-1)$
313	10016	51	$(19, 97, 51, 1, 252)$	$0.177(-1)$
619	19808	104	$(1, 104, 427, 141, 293)$	$0.781(-2)$
1249	39968	165	$(310, 996, 1, 721, 165)$	$0.258(-2)$
2503	80096	792	$(1514, 1951, 792, 151, 1)$	$0.962(-3)$
5003	160096	380	$(1687, 4099, 4316, 1, 380)$	$0.367(-3)$
10007	320224	1927	$(732, 1927, 5453, 9584, 1)$	$0.137(-3)$

Table A.5 Recommended choices of **z** for $s = 6$

m	$2^s m$	ℓ	**z**	$P_2(Q_s)$
79	5056	27	$(8, 27, 1, 18, 12, 58)$	$0.279(0)$
157	10048	18	$(18, 10, 100, 23, 73, 1)$	$0.126(0)$
313	20032	80	$(194, 140, 183, 245, 1, 80)$	$0.510(-1)$
619	39616	102	$(295, 102, 242, 543, 500, 1)$	$0.229(-1)$
1249	79936	175	$(1, 175, 649, 440, 1165, 288)$	$0.950(-2)$
2503	160192	253	$(253, 2009, 1, 1434, 1393, 2370)$	$0.367(-2)$
5003	320192	162	$(4718, 4538, 1, 1229, 162, 3981)$	$0.137(-2)$
10007	640448	2286	$(3189, 1, 2142, 2286, 6064, 4958)$	$0.548(-3)$

Table A.6 Recommended choices of $\mathbf{z}$ for $s = 7$

m	$2^s m$	ℓ	$\mathbf{z}$	$P_2(Q_s)$
41	5248	13	$(1, 25, 5, 13, 2, 24, 38)$	$0.134(1)$
79	10112	27	$(12, 65, 27, 1, 18, 8, 58)$	$0.611(0)$
157	20096	11	$(121, 40, 126, 75, 1, 11, 130)$	$0.291(0)$
313	40064	70	$(205, 70, 265, 1, 83, 176, 113)$	$0.128(0)$
619	79232	161	$(542, 289, 161, 1, 71, 358, 602)$	$0.570(-1)$
1249	159872	303	$(993, 1, 303, 632,$ $1119, 399, 578)$	$0.239(-1)$
2503	320384	306	$(924, 306, 2408, 1025,$ $1, 1868, 775)$	$0.100(-1)$
5003	640384	2037	$(1, 1336, 2037, 1882,$ $4803, 2846, 3828)$	$0.407(-2)$
10007	1280896	2286	$(1, 2286, 2142, 3189,$ $4958, 6064, 2609)$	$0.165(-2)$

Table A.7 Recommended choices of $\mathbf{z}$ for $s = 8$

m	$2^s m$	ℓ	$\mathbf{z}$	$P_2(Q_s)$
19	4864	2	$(16, 14, 7, 4, 8, 13, 2, 1)$	$0.585(1)$
41	10496	13	$(1, 25, 5, 26, 13, 2, 24, 38)$	$0.265(1)$
79	20224	27	$(1, 17, 27, 18, 12, 8, 65, 58)$	$0.127(1)$
157	40192	11	$(75, 1, 40, 130, 126, 11, 121, 17)$	$0.598(0)$
313	80128	70	$(85, 83, 205, 70, 1, 265, 176, 113)$	$0.279(0)$
619	158464	161	$(358, 71, 161, 289,$ $104, 602, 542, 1)$	$0.127(0)$
1249	319744	155	$(294, 903, 30, 606,$ $806, 255, 155, 1)$	$0.565(-1)$
2503	640768	153	$(1994, 153, 1, 2287,$ $882, 1602, 2315, 2219)$	$0.248(-1)$
5003	1280768	137	$(4789, 4814, 700, 843,$ $137, 4125, 1, 3760)$	$0.111(-1)$
10007	2561792	373	$(8310, 1, 8822, 9038,$ $373, 3245, 9545, 7467)$	$0.464(-2)$

Table A.8 Recommended choices of $\mathbf{z}$ for $s = 9$

m	$2^s m$	ℓ	$\mathbf{z}$	$P_2(Q_s)$
11	5632	2	$(10, 8, 3, 5, 4, 9, 7, 2, 1)$	0.195(2)
19	9728	2	$(8, 9, 13, 14, 1, 16, 2, 7, 4)$	0.111(2)
41	20992	13	$(13, 2, 24, 26, 10, 1, 5, 25, 38)$	0.504(1)
79	40448	27	$(58, 27, 1, 17, 8, 12, 65, 18, 64)$	0.250(1)
157	80384	11	$(1, 17, 75, 121, 126,$ $130, 40, 11, 30)$	0.121(1)
313	160256	93	$(93, 1, 195, 79, 294,$ $305, 198, 148, 260)$	0.573(0)
619	316928	161	$(602, 289, 31, 104, 542,$ $1, 161, 358, 71)$	0.267(0)
1249	639488	18	$(1080, 18, 324, 60, 836,$ $705, 200, 1102, 1)$	0.126(0)
2503	1281536	153	$(882, 153, 2219, 1, 1994,$ $2287, 2315, 1602, 1272)$	0.551(−1)
5003	2561536	657	$(2419, 657, 3341, 4547, 588,$ $3723, 1085, 1391, 1)$	0.259(−1)
10007	5123584	563	$(8561, 1, 563, 8723, 7508,$ $7619, 4050, 6752, 6501)$	0.115(−1)

Table A.9 Recommended choices of $\mathbf{z}$ for $s = 10$

m	$2^s m$	ℓ	$\mathbf{z}$	$P_2(Q_s)$
5	5120	2	$(2, 2, 3, 3, 1, 4, 4, 1, 2, 1)$	0.799(2)
11	11264	2	$(10, 3, 8, 5, 6, 7, 9, 2, 4, 1)$	0.360(2)
19	19456	2	$(8, 9, 13, 14, 16, 18, 1, 2, 4, 7)$	0.206(2)
41	41984	13	$(7, 2, 13, 24, 26, 1, 5, 10, 25, 38)$	0.943(1)
79	80896	27	$(1, 58, 17, 27, 8,$ $12, 69, 65, 18, 64)$	0.478(1)
157	160768	36	$(30, 138, 115, 101, 58,$ $(27, 36, 40, 1, 25)$	0.234(1)
313	320512	62	$(301, 135, 20, 299, 195,$ $232, 88, 71, 62, 1)$	0.113(1)

Continued on next page

m	$2^s m$	ℓ	$\mathbf{z}$	$P_2(Q_s)$
619	633856	106	$(69, 505, 588, 60, 170,$ $94, 106, 296, 426, 1)$	$0.542(0)$
1249	1278976	184	$(1026, 1241, 133, 741, 203,$ $1131, 770, 543, 1, 184)$	$0.257(0)$
2503	2563072	1019	$(1019, 2119, 1, 71, 1284,$ $2265, 2282, 269, 1675, 1830)$	$0.119(0)$
5003	5123072	189	$(4326, 1385, 4709, 4470, 700,$ $1609, 189, 2125, 1, 2222)$	$0.545(-1)$
10007	10247168	1047	$(8175, 3879, 1, 9914, 3240,$ $1047, 7979, 8478, 5446, 2699)$	$0.257(-1)$

Table A.10 Recommended choices of $\mathbf{z}$ for $s = 11$

m	$2^s m$	ℓ	$\mathbf{z}$	$P_2(Q_s)$
3	6144	1	$(1, 1, 1, 1, 1, 1, 1, 1, 1, 1, 1)$	$0.245(3)$
5	10240	2	$(1, 4, 4, 1, 4, 2, 3, 3, 2, 2, 1)$	$0.146(3)$
11	22528	2	$(10, 1, 3, 1, 5, 6, 9, 7, 2, 4, 8)$	$0.662(2)$
19	38912	4	$(1, 4, 4, 1, 9, 11, 16, 5, 17, 6, 7)$	$0.381(2)$
41	83968	13	$(2, 7, 13, 24, 5, 26,$ $10, 1, 9, 25, 38)$	$0.175(2)$
79	161792	8	$(67, 18, 64, 65, 38, 46,$ $8, 52, 1, 22, 62)$	$0.894(1)$
157	321536	36	$(47, 30, 25, 101, 36, 58,$ $138, 40, 1, 27, 115)$	$0.442(1)$
313	641024	15	$(133, 1, 242, 37, 225, 245,$ $15, 232, 187, 117, 301)$	$0.217(1)$
619	1267712	106	$(588, 296, 505, 69, 426, 60,$ $94, 170, 106, 428, 1)$	$0.106(1)$
1249	2557952	113	$(302, 39, 660, 113, 575, 27,$ $279, 403, 889, 1, 553)$	$0.508(0)$
2503	5126144	29	$(1, 951, 46, 1759, 1862, 1269,$ $1435, 841, 29, 1567, 389)$	$0.241(0)$
5003	10246144	189	$(700, 1609, 1, 4709, 1385, 4326,$ $3921, 2125, 189, 4470, 2222)$	$0.114(0)$
10007	20494336	2499	$(1, 8722, 1032, 761, 2499, 4806,$ $409, 2801, 1377, 7169, 633)$	$0.531(-1)$

Table A.11 Recommended choices of z for $s = 12$

m	$2^s m$	ℓ	z	$P_2(Q_s)$
3	12288	1	$(1,1,1,1,1,1,$ $1,1,1,1,1,1)$	0.447(3)
5	20480	2	$(2,3,3,2,2,4,$ $1,1,4,4,3,1)$	0.268(3)
11	45056	3	$(4,1,1,1,9,9,$ $5,3,3,5,4,3)$	0.121(3)
19	77824	3	$(16,15,6,3,7,2,$ $10,18,1,8,5,9)$	0.700(2)
41	167936	13	$(1,10,38,5,7,9,$ $24,26,2,35,25,13)$	0.322(2)
79	323584	8	$(67,64,18,21,65,38,$ $46,8,52,1,22,62)$	0.166(2)
157	643072	36	$(138,58,25,27,30,122,$ $115,40,47,36,101,1)$	0.824(1)
313	1282048	15	$(242,301,37,1,15,187,$ $245,117,232,225,190,133)$	0.408(1)
619	2535424	80	$(138,1,210,141,517,506,$ $(245,411,87,319,151,80)$	0.202(1)
1249	5115904	467	$(1,598,587,763,595,739,$ $135,558,467,389,794,356)$	0.974(0)
2503	10252288	29	$(1334,1759,841,1,1862,1435,$ $951,1269,1567,29,46,389)$	0.468(0)
5003	20492288	331	$(4698,4604,331,1042,$ $1375,3012,4875,1,$ $2659,4498,4855,2947)$	0.227(0)
10007	40988672	745	$(4640,745,4415,6879,$ $1,4543,4385,7125,$ $4778,6237,1271,2169)$	0.108(0)

Appendix B

Tables of z for nonperiodic integrands

Here we give tables of $\mathbf{z}$ obtained by finding a value of ℓ in

$$\mathbf{z}(\ell) = (1, \ell, \ell^2 \bmod N, \ldots, \ell^{s-1} \bmod N), \qquad 1 \leqslant \ell \leqslant \lfloor N/2 \rfloor,$$

such that the vertex variance

$$v_B = \sum_{i_s=0}^{1} \cdots \sum_{i_1=0}^{1} W_{i_1 \cdots i_s}^2$$

is minimized. Here $W_{i_1 \cdots i_s}$, for $i_1, \ldots, i_s = 0$ or 1, are the weights in the 'optimal' quadrature rule

$$Bf := \sum_{i_s=0}^{1} \cdots \sum_{i_1=0}^{1} W_{i_1 \cdots i_s} f(i_1, \ldots, i_s) + \frac{1}{N} \sum_{j=1}^{N-1} f\left(\left\{\frac{j}{N}\mathbf{z}(\ell)\right\}\right).$$

Further details may be found in Sections 8.2 and 8.3.

Table B.1 Recommended choices of ℓ for $s = 2$

N	ℓ	$N^2 v_B$	P_2
79	15	0.250(0)	0.276(−1)
157	28	0.250(0)	0.865(−2)
313	25	0.250(0)	0.749(−2)
619	74	0.250(0)	0.924(−3)
1249	585	0.250(0)	0.245(−3)
2503	340	0.250(0)	0.624(−4)
5003	2318	0.250(0)	0.202(−4)
10007	4722	0.250(0)	0.507(−5)

Table B.2 Recommended choices of ℓ for $s = 3$

N	ℓ	$N^2 v_B$	P_2
79	23	0.166(0)	0.222(1)
157	22	0.145(0)	0.169(0)
313	51	0.131(0)	0.625(−1)
619	239	0.149(0)	0.228(−1)
1249	507	0.130(0)	0.164(−1)
2503	521	0.140(0)	0.199(−2)
5003	1086	0.126(0)	0.193(−2)
10007	4073	0.126(0)	0.106(−1)

Table B.3 Recommended choices of ℓ for $s = 4$

N	ℓ	$N^2 v_B$	P_2
79	29	0.333(0)	0.673(1)
157	22	0.229(0)	0.128(1)
313	51	0.445(0)	0.512(0)
619	173	0.531(0)	0.218(1)
1249	338	0.169(0)	0.633(−1)
2503	183	0.377(0)	0.814(−1)
5003	1033	0.468(0)	0.155(−1)
10007	1481	0.224(0)	0.556(−2)

Table B.4 Recommended choices of ℓ for $s = 5$

N	ℓ	$N^2 v_B$	P_2
79	15	0.466(0)	0.171(2)
157	22	0.571(0)	0.902(1)
313	38	0.110(1)	0.345(1)
619	132	0.149(1)	0.174(1)
1249	377	0.165(1)	0.747(0)
2503	814	0.838(0)	0.419(0)
5003	2036	0.161(1)	0.151(0)
10007	1481	0.338(1)	0.599(−1)

Table B.5 Recommended choices of ℓ for $s = 6$

N	ℓ	$N^2 v_B$	P_2
79	15	0.804(0)	0.774(2)
157	22	0.813(0)	0.402(2)
313	59	0.140(1)	0.195(2)
619	132	0.231(1)	0.902(1)
1249	143	0.329(1)	0.372(1)
2503	347	0.237(1)	0.222(1)
5003	708	0.643(1)	0.781(0)
10007	357	0.877(1)	0.157(1)

Table B.6 Recommended choices of ℓ for $s = 7$

N	ℓ	$N^2 v_B$	P_2
79	6	0.844(0)	0.334(3)
157	42	0.136(1)	0.171(3)
313	59	0.179(1)	0.852(2)
619	53	0.377(1)	0.409(2)
1249	131	0.501(1)	0.203(2)
2503	530	0.648(1)	0.944(1)
5003	687	0.970(1)	0.456(1)
10007	1855	0.146(2)	0.247(1)

Table B.7 Recommended choices of ℓ for $s = 8$

N	ℓ	$N^2 v_B$	P_2
79	15	0.847(0)	0.145(4)
157	42	0.128(1)	0.736(3)
313	59	0.187(1)	0.365(3)
619	53	0.337(1)	0.185(3)
1249	235	0.514(1)	0.900(2)
2503	223	0.863(1)	0.475(2)
5003	687	0.117(2)	0.213(2)
10007	1855	0.227(2)	0.108(2)

Table B.8 Recommended choices of ℓ for $s = 9$

N	ℓ	$N^2 v_B$	P_2
79	28	0.698(0)	0.623(4)
157	36	0.118(1)	0.313(4)
313	59	0.179(1)	0.156(4)
619	53	0.308(1)	0.794(3)
1249	235	0.450(1)	0.390(3)
2503	223	0.810(1)	0.203(3)
5003	687	0.153(2)	0.969(2)
10007	2521	0.229(2)	0.482(2)

Table B.9 Recommended choices of ℓ for $s = 10$

N	ℓ	$N^2 v_B$	P_2
79	28	0.539(0)	0.267(5)
157	36	0.966(0)	0.134(5)
313	59	0.167(1)	0.673(4)
619	53	0.261(1)	0.341(4)
1249	235	0.423(1)	0.168(4)
2503	372	0.816(1)	0.836(3)
5003	1292	0.156(2)	0.417(3)
10007	4474	0.244(2)	0.206(3)

Table B.10 Recommended choices of ℓ for $s = 11$

N	ℓ	$N^2 v_B$	P_2
79	28	0.402(0)	0.115(6)
157	36	0.758(0)	0.577(5)
313	62	0.149(1)	0.289(5)
619	40	0.277(1)	0.146(5)
1249	235	0.439(1)	0.724(4)
2503	372	0.805(1)	0.361(4)
5003	1180	0.145(2)	0.180(4)
10007	4474	0.235(2)	0.900(3)

Table B.11 Recommended choices of ℓ for $s = 12$

N	ℓ	$N^2 v_B$	P_2
79	28	0.299(0)	0.493(6)
157	66	0.561(0)	0.247(6)
313	62	0.114(1)	0.124(6)
619	40	0.217(1)	0.628(5)
1249	328	0.395(1)	0.311(5)
2503	140	0.678(1)	0.155(5)
5003	201	0.123(2)	0.776(4)
10007	3708	0.205(2)	0.388(4)

Appendix C

Bernoulli polynomials

The Bernoulli polynomials are defined (Gradshteyn and Ryzhik 1980) by the generating function

$$\frac{te^{xt}}{e^t - 1} = \sum_{n=0}^{\infty} B_n(x)\frac{t^n}{n!}, \qquad |t| < 2\pi. \tag{C.1}$$

From this it follows easily that B_n is a polynomial of degree n. Useful special cases are

$$
\begin{aligned}
B_0(x) &= 1, \\
B_1(x) &= x - \frac{1}{2}, \\
B_2(x) &= x^2 - x + \frac{1}{6}.
\end{aligned}
$$

Bernoulli polynomials have a number of properties that make them useful in numerical analysis: they behave simply under differentiation and integration, and possess simple Fourier expansions. Their properties are even nicer if we depart from the classical normalization and introduce the renormalized Bernoulli polynomials

$$b_n(x) = \frac{B_n(x)}{n!}.$$

The generating function definition (C.1) then becomes

$$\frac{te^{xt}}{e^t - 1} = \sum_{n=0}^{\infty} b_n(x)t^n. \tag{C.2}$$

Integration of this from 0 to 1 yields

$$\int_0^1 b_n(x)dx = \begin{cases} 1 & \text{if } n = 0, \\ 0 & \text{if } n \geqslant 1, \end{cases}$$

while differentiation of (C.2) yields

$$b_n'(x) = b_{n-1}(x), \qquad n \geqslant 1.$$

Repeated application of these two relations yields

$$\int_0^1 b_n^{(m)}(x)\,dx = \begin{cases} 1 & \text{if } m = n, \\ 0 & \text{if } m \neq n. \end{cases}$$

This property is useful in the present text, and also underlies the role of Bernoulli polynomials in the Euler–Maclaurin expansion.

Multiplication of (C.2) by $e^{-2\pi i h x}$ and integration from 0 to 1 yields

$$\int_0^1 b_n(x)e^{-2\pi i h x}\,dx = -\frac{1}{(2\pi i h)^n}, \qquad n \geqslant 1, \qquad h \neq 0,$$

from which follows the Fourier series representation of the Bernoulli polynomials,

$$b_n(x) = -\sum_{h \neq 0} \frac{1}{(2\pi i h)^n} e^{2\pi i h x}, \qquad n \geqslant 1, \qquad x \in (0,1). \tag{C.3}$$

In the case where $n = 1$ the convergence in (C.3) is only conditional, and the right-hand side must be understood as the limit of the symmetric partial sums. In all other cases the pointwise convergence in (C.3) extends to the closed interval $[0,1]$.

References

Babenko, K.I. (1960*a*). Approximation of periodic functions of many variables by trigonometric polynomials. *Doklady Akademii Nauk SSSR*, **132**, 247–50 (Russian). English transl.: *Soviet Mathematics Doklady*, **1**, 513–6.

Babenko, K.I. (1960*b*). Approximation by trigonometric polynomials in a certain class of periodic functions of several variables. *Doklady Akademii Nauk SSSR*, **132**, 982–5 (Russian). English transl.: *Soviet Mathematics Doklady*, **1**, 672–5.

Bahvalov, N.S. (1959). On approximate calculation of multiple integrals. *Vestnik Moskovskogo Universiteta, Seriya Matematiki, Mehaniki, Astronomii, Fiziki, Himii*, **4**, 3–18 (Russian).

Beckers, M. (1992). Numerical integration in high dimensions. Unpublished Ph.D. thesis. Katholieke Universiteit, Leuven.

Beckers, M. and Cools, R. (1993). A relation between cubature formulae of trigonometric degree and lattice rules. In *Numerical integration IV*, International Series of Numerical Mathematics, No. 112 (eds. H. Brass and G. Hämmerlin), pp. 13–24. Birkhäuser, Basel.

Beckers, M. and Haegemans, A. (1992*a*). Transformation of integrands for lattice rules. In *Numerical integration: recent developments, software and applications* (eds. T.O. Espelid and A.C. Genz), pp. 329–40. Kluwer Academic Publishers, Dordrecht.

Beckers, M. and Haegemans, A. (1992*b*). *Comparison of numerical integration techniques for multivariate normal integrals*, Preprint. Department of Computer Science, Katholieke Universiteit, Leuven.

Berntsen J. and Espelid, T.O. (1991). Error estimation in automatic quadrature routines. *ACM Transactions on Mathematical Software*, **17**, 233–52.

Berntsen J. and Espelid, T.O. (1992). Algorithm 706: DCUTRI: An algorithm for adaptive cubature over a collection of triangles. *ACM Transactions on Mathematical Software*, **18**, 329–42.

Berntsen J., Espelid, T.O., and Genz, A.C. (1991*a*). An adaptive algorithm for the approximate calculation of multiple integrals. *ACM Transactions on Mathematical Software*, **17**, 437–51.

Berntsen J., Espelid, T.O., and Genz, A.C. (1991*b*). Algorithm 698: DCUHRE—An adaptive multidimensional integration routine for a vector of integrals. *ACM Transactions on Mathematical Software*, **17**, 452–6.

Berntsen J., Cools, R., and Espelid, T.O. (1993). Algorithm 720: An algorithm for adaptive cubature over a collection of 3-dimensional simplices. *ACM Transactions on Mathematical Software*, **19**, 320–32.

Cassels, J.W.S. (1971). *An introduction to the geometry of numbers.* Springer Verlag, Berlin.

Conroy, H. (1967). Molecular Schrödinger equation VIII: A new method for the evaluation of multidimensional integrals. *The Journal of Chemical Physics*, **47**, 5307–18.

Cools, R. (1992). A survey of methods for constructing cubature formulae. In *Numerical integration: recent developments, software and applications* (eds. T.O. Espelid and A.C. Genz), pp. 1–24, Kluwer Academic Publishers, Dordrecht.

Cools, R. and Rabinowitz, P. (1993). Monomial cubature rules since 'Stroud': a compilation. *Journal of Computational and Applied Mathematics*, **43**, 309–26.

Cools, R. and Sloan, I.H. (1993). *Minimal cubature formulae of trigonometric degree*, Preprint AMR93/5. Department of Applied Mathematics, University of New South Wales, Sydney.

Cranley, R. and Patterson, T.N.L. (1976). Randomization of number theoretic methods for multiple integration. *SIAM Journal on Numerical Analysis*, **13**, 904–14.

Davis, P.J. and Rabinowitz, P. (1984). *Methods of numerical integration* (2nd edn). Academic Press, London.

de Doncker, E. and Robinson, I.G. (1984*a*). An algorithm for automatic integration over a triangle using nonlinear extrapolation. *ACM Transactions on Mathematical Software*, **10**, 1–16.

de Doncker, E. and Robinson, I.G. (1984*b*). Algorithm 612. TRIEX: Integration over a TRIangle using nonlinear EXtrapolation. *ACM Transactions on Mathematical Software*, **10**, 17–22.

Disney, S.A.R. (1990). Error bounds for rank 1 lattice quadrature rules modulo composites. *Monatshefte für Mathematik*, **110**, 89–100.

Disney, S.A.R. and Sloan, I.H. (1991). Error bounds for the method of good lattice points. *Mathematics of Computation*, **56**, 257–66.

Disney, S.A.R. and Sloan, I.H. (1992). Lattice integration rules of maximal rank formed by copying rank 1 rules. *SIAM Journal on Numerical Analysis*, **29**, 566–77.

Engels, H. (1980). *Numerical quadrature and cubature.* Academic Press, London.

Frolov, K.K. (1977). On the connection between quadrature formulas and sublattices of the lattice of integral vectors. *Doklady Akademii Nauk SSSR*, **232**, 40–3 (Russian). English transl.: *Soviet Mathematics Doklady*, **18**, 37–41.

Genz, A.C. (1984). Testing multidimensional integration routines. In *Tools, methods, and languages for scientific and engineering computation* (eds. B. Ford, J.C. Rault, and F. Thomasset), pp. 81–94. North-Holland, Amsterdam.

Genz, A.C. (1987). A package for testing multiple integration subroutines. In *Numerical integration: recent developments, software and applications* (eds. P. Keast and G. Fairweather), pp. 337–40. D. Reidel Publishing, Dordrecht.

Genz, A.C. and Malik, A.A. (1980). An adaptive algorithm for numerical integration over an N-dimensional rectangular region. *Journal of Computational and Applied Mathematics*, **6**, 295–302.

Gradshteyn, I.S. and Ryzhik, I.M. (1980), *Tables of integrals, series, and products* (4th edn). Academic Press, New York.

Haber, S. (1970). Numerical evaluation of multiple integrals. *SIAM Review*, **12**, 481–526.

Haber, S. (1972). Experiments on optimal coefficients. In *Applications of number theory to numerical analysis* (ed. S.K. Zaremba), pp. 11–37. Academic Press, New York.

Haber, S. (1983). Parameters for integrating periodic functions of several variables. *Mathematics of Computation*, **41**, 115–29.

Haegemans, A. (1977). Algorithm 34: an algorithm for the automatic integration over a triangle. *Computing*, **19**, 179–87.

Halton, J.H. (1960). On the efficiency of certain quasi-random sequences of points in evaluating multidimensional integrals. *Numerische Mathematik*, **2**, 84–90.

Hammersley, J.M. (1960). Monte Carlo methods for solving multivariate problems. *Annals of the New York Academy of Sciences*, **86**, 844–74.

Hammersley, J.M. and Handscomb D.C. (1964). *Monte Carlo methods*. Methuen, London.

Hardy, G.H., Littlewood, J.E., and Pólya G. (1934). *Inequalities*. Cambridge University Press, Cambridge.

Hlawka, E. (1961). Funktionen von beschränkter Variation in der Theorie der Gleichverteilung. *Annali di Matematica Pura ed Applicata*, **54**, 325–34.

Hlawka, E. (1962). Zur angenäherten Berechnung mehrfacher Integrale. *Monatshefte für Mathematik*, **66**, 140–51.

Hua, L.K. and Wang, Y. (1981). *Applications of number theory to numerical analysis*. Springer Verlag, Berlin; Science Press, Beijing.

Joe, S. (1989). The generation of lattice points for numerical multiple integration. *Journal of Computational and Applied Mathematics*, **26**, 327–31.

Joe, S. (1990*a*). Randomization of lattice rules for numerical multiple integration. *Journal of Computational and Applied Mathematics*, **31**, 299–304.

Joe, S. (1990*b*). *A curiosity arising from searches for good lattice points*, Preprint AM90/8. Department of Applied Mathematics, University of New South Wales, Sydney.

Joe, S. and Disney, S.A.R. (1993). Intermediate rank lattice rules for multidimensional integration. *SIAM Journal on Numerical Analysis*, **30**, 569–82.

Joe, S. and Hunt, D.C. (1992). The number of lattice rules having given invariants. *Bulletin of the Australian Mathematical Society*, **46**, 479–95.

Joe, S. and Sloan, I.H. (1992*a*). Embedded lattice rules for multidimensional integration. *SIAM Journal on Numerical Analysis*, **29**, 1119–35.

Joe, S. and Sloan, I.H. (1992*b*). On computing the lattice rule criterion R. *Mathematics of Computation*, **59**, 557–68.

Joe, S. and Sloan, I.H. (1993). Implementation of a lattice method for numerical multiple integration. *ACM Transactions on Mathematical Software*, **19**, 523–45.

Kalos, M.H. and Whitlock, P.A. (1986). *Monte Carlo methods Volume I: Basics*. Wiley, New York.

Knuth, D.E. (1973). *The art of computer programming Volume 2: Seminumerical algorithms* (2nd edn). Addison-Wesley, Reading.

Korobov, N.M. (1959). The approximate computation of multiple integrals. *Doklady Akademii Nauk SSSR*, **124**, 1207–10 (Russian).

Korobov, N.M. (1960). Properties and calculation of optimal coefficients. *Doklady Akademii Nauk SSSR*, **132**, 1009–12 (Russian). English transl.: *Soviet Mathematics Doklady*, **1**, 696–700.

Korobov, N.M. (1963). *Number-theoretic methods in approximate analysis*. Fizmatgiz, Moscow (Russian).

Larcher, G. and Niederreiter, H. (1989). Optimal coefficients modulo prime powers in the three-dimensional case. *Annali di Matematica Pura ed Applicata*, **155**, 299–315.

Laurie, D.P. (1982). Algorithm 584. CUBTRI: Automatic cubature over a triangle. *ACM Transactions on Mathematical Software*, **8**, 210–8.

Lautrup, B. (1971). An adaptive multidimensional integration procedure. In *Proceedings of the Second Colloquium on Advanced Computing Methods in Theoretical Physics*, pp. I 57–82. C.N.R.S., Marseille.

Ledermann, W. (1964). *Introduction to the theory of finite groups* (5th edn). Oliver and Boyd, Edinburgh.

Lindgren, B.W. (1976). *Statistical Theory* (3rd edn). Collier Macmillan, New York.

Lyness, J.N. (1988). Some comments on quadrature rule construction criteria. In *Numerical integration III*, International Series of Numerical Mathematics, No. 85 (eds. H. Brass and G. Hämmerlin), pp. 117–29. Birkhäuser, Basel.

Lyness, J.N. (1989). An introduction to lattice rules and their generator matrices. *IMA Journal of Numerical Analysis*, **9**, 405–19.

Lyness, J.N. (1993). The canonical forms of a lattice rule. In *Numerical integration IV*, International Series of Numerical Mathematics, No. 112 (eds. H. Brass and G. Hämmerlin), pp. 225–40. Birkhäuser, Basel.

Lyness, J.N. and Sloan, I.H. (1989). Some properties of rank-2 lattice rules. *Mathematics of Computation*, **53**, 627–37.

Lyness, J.N. and Sørevik, T. (1989). The number of lattice rules. *BIT*, **29**, 527–34.

Maisonneuve, D. (1972). Recherche et utilisation des 'bons trellis.' Programmation et résultats numériques. In *Applications of number theory to numerical analysis* (ed. S.K. Zaremba), pp. 121–201. Academic Press, New York.

Metropolis, N.C. and Ulam, S.M. (1949). The Monte Carlo method. *Journal of the American Statistical Association*, **44**, 335–41.

Mysovskikh, I.P. (1987). On cubature formulas that are exact for trigonometric polynomials. *Doklady Akademii Nauk SSSR*, **296**, 28–31 (Russian). English transl.: *Soviet Mathematics Doklady*, **36**, 229–32.

Newman, M. (1972). *Integer matrices*. Academic Press, New York.

Niederreiter, H. (1978*a*). Quasi-Monte Carlo methods and pseudo-random numbers. *Bulletin of the American Mathematical Society*, **84**, 957–1041.

Niederreiter, H. (1978*b*). Existence of good lattice points in the sense of Hlawka. *Monatshefte für Mathematik*, **86**, 203–19.

Niederreiter, H. (1988). Quasi-Monte Carlo methods for multidimensional numerical integration. In *Numerical integration III*, International Series of Numerical Mathematics, No. 85 (eds. H. Brass and G. Hämmerlin), pp. 157–71. Birkhäuser, Basel.

Niederreiter, H. (1992*a*). The existence of efficient lattice rules for multidimensional numerical integration. *Mathematics of Computation*, **58**, 305–14 and S7–S16.

Niederreiter, H. (1992*b*). *Random number generation and quasi-Monte Carlo methods*. Society for Industrial and Applied Mathematics, Philadelphia.

Niederreiter, H. (1992*c*). Existence theorems for efficient lattice rules. In *Numerical integration: recent developments, software and applications* (eds. T.O. Espelid and A.C. Genz), pp. 71–80. Kluwer Academic Publishers, Dordrecht.

Niederreiter, H. (1993). Improved error bounds for lattice rules. *Journal of Complexity*, **9**, 60–75.

Niederreiter, H. and Sloan, I.H. (1990). Lattice rules for multiple integration and discrepancy. *Mathematics of Computation*, **54**, 303–12.

Niederreiter, H. and Sloan, I.H. (1993). Quasi-Monte Carlo methods with modified vertex weights. In *Numerical integration IV*, International Series of Numerical Mathematics, No. 112 (eds. H. Brass and G. Hämmerlin), pp. 253–65. Birkhäuser, Basel.

Niederreiter, H. and Sloan, I.H. Integration of nonperiodic functions of two variables by Fibonacci lattice rules. *Journal of Computational and Applied Mathematics*. (In press.)

Niven, I. and Zuckerman, H.S. (1980). *An introduction to the theory of numbers* (4th edn). Wiley, New York.

Numerical Algorithms Group (1991). *NAG Fortran Library, Mark 15*. Numerical Algorithms Group Ltd, Oxford.

Price, J.F. and Sloan, I.H. (1992). Pointwise convergence of multiple Fourier series: sufficient conditions and an application to numerical integration. *Journal of Mathematical Analysis and Applications*, **169**, 140–56.

Sag, T.W. and Szekeres, G. (1964). Numerical evaluation of high-dimensional integrals. *Mathematics of Computation*, **18**, 245–53.

Saltykov, A.I. (1963). Tables for the computation of multiple integrals using the method of optimal coefficients. *Zhurnal Vychislitel'noĭ Matematiki i Matematicheskoĭ Fiziki*, **3**, 181–6 (Russian). English transl.: *U.S.S.R. Computational Mathematics and Mathematical Physics*, **3**, 235–42.

Šarygin, I.F. (1963). A lower estimate for the error of quadrature formulas for certain classes of functions. *Zhurnal Vychislitel'noĭ Matematiki i Matematicheskoĭ Fiziki*, **3**, 370–6 (Russian). English transl.: *U.S.S.R. Computational Mathematics and Mathematical Physics*, **3**, 489–97.

Schrage, L. (1979). A more portable Fortran random number generator. *ACM Transactions on Mathematical Software*, **5**, 132–8.

Sidi, A. (1993). A new variable transformation for numerical integration. In *Numerical integration IV*, International Series of Numerical Mathematics, No. 112 (eds. H. Brass and G. Hämmerlin), pp. 359–73. Birkhäuser, Basel.

Sloan, I.H. (1985). Lattice methods for multiple integration. *Journal of Computational and Applied Mathematics*, **12 & 13**, 131–43.

Sloan, I.H. (1992). Numerical integration in high dimensions—the lattice rule approach. In *Numerical integration: recent developments, software and applications* (eds. T.O. Espelid and A.C. Genz), pp. 55–69. Kluwer Academic Publishers, Dordrecht.

Sloan, I.H. and Kachoyan, P.J. (1984). Lattices for multiple integration. In *Mathematical programming and numerical analysis workshop*, Proceedings of the Centre for Mathematical Analysis, No. 6, pp. 147–65. Australian National University, Canberra.

Sloan, I.H. and Kachoyan, P.J. (1987). Lattice methods for multiple integration: theory, error analysis and examples. *SIAM Journal on Numerical Analysis*, **24**, 116–28.

Sloan, I.H. and Lyness, J.N. (1989). The representation of lattice quadrature rules as multiple sums. *Mathematics of Computation*, **52**, 81–94.

Sloan, I.H. and Lyness, J.N. (1990). Lattice rules: projection regularity and unique representations. *Mathematics of Computation*, **54**, 649–60.

Sloan, I.H. and Osborn, T.R. (1987). Multiple integration over bounded and unbounded regions. *Journal of Computational and Applied Mathematics*, **17**, 181–96.

Sloan, I.H. and Walsh, L. (1990). A computer search of rank 2 lattice rules for multidimensional quadrature. *Mathematics of Computation*, **54**, 281–302.

Sloane, N.J.A. (1981). Tables of sphere packings and spherical codes. *IEEE Transactions on Information Theory*, **27**, 327–38.

Sobolev S.L. (1962). Cubature formulas on the sphere invariant under finite groups of rotations. *Doklady Akademii Nauk SSSR*, **146**, 310–3 (Russian). English transl.: *Soviet Mathematics Doklady*, **3**, 1307–10.

Stein, E.M. and Weiss, G. (1971). *Introduction to Fourier analysis on Euclidean spaces*. Princeton University Press, Princeton.

Stroud, A.H. (1971). *Approximate calculation of multiple integrals*. Prentice-Hall, Englewood Cliffs.

Sugihara, M. (1987). Method of good matrices for multidimensional numerical integrations—An extension of the method of good lattice points. *Journal of Computational and Applied Mathematics*, **17**, 197–213.

Van Dooren, P. and De Ridder, L. (1976). An adaptive algorithm for numerical integration over an n-dimensional cube. *Journal of Computational and Applied Mathematics*, **2**, 207–17.

Zaremba, S.K. (1966). Good lattice points, discrepancy, and numerical integration. *Annali di Matematica Pura ed Applicata*, **73**, 293–317.

Zaremba, S.K. (1968). Some applications of multidimensional integration by parts. *Annales Polonici Mathematici*, **21**, 85–96.

Zaremba, S.K. (1972). La méthode des 'bons trellis' pour le calcul des intégrales multiples. In *Applications of number theory to numerical analysis* (ed. S.K. Zaremba), pp. 39–116. Academic Press, New York.

Index